Advanced Model Predictive Control

Advanced Model Predictive Control

Editor

Bianca Lupei

Advanced Model Predictive Control

Edited by **Bianca Lupei**

ISBN: 978-1-68117-205-7
Library of Congress Control Number: 2016934752

Notice

Preface

Model predictive control is an advanced method of process control that has been in use in the process industries in chemical plants and oil refineries since the 1980s. In recent years it has also been used in power system balancing models. Model predictive controllers rely on dynamic models of the process, most often linear empirical models obtained by system identification. The main advantage of model predictive control is the fact that it allows the current timeslot to be optimized, while keeping future timeslots in account. This is achieved by optimizing a finite time-horizon, but only implementing the current timeslot. Model predictive control has the ability to anticipate future events and can take control actions accordingly. MPC models predict the change in the dependent variables of the modelled system that will be caused by changes in the independent variables. In a chemical process, independent variables that can be adjusted by the controller are often either the setpoints of regulatory PID controllers or the final control element. Independent variables that cannot be adjusted by the controller are used as disturbances. Dependent variables in these processes are other measurements that represent either control objectives or process constraints.

The book entitled Advanced Model Predictive Control is intended to present the readers the recent achievements in this field. The book also delivers applications of MPC in modern industry and effective commercial software for MPC is familiarized.

Table of Contents

CHAPTER 1

Efficient Nonlinear Model Predictive Control for Affine System

Tao ZHENG and Wei CHEN

Hefei University of Technology, China

1. INTRODUCTION

Model predictive control (MPC) refers to the class of computer control algorithms in which a dynamic process model is used to predict and optimize process performance. Since its lower request of modeling accuracy and robustness to complicated process plants, MPC for linear systems has been widely accepted in the process industry and many other fields. But for highly nonlinear processes, or for some moderately nonlinear processes with large operating regions, linear MPC is often inefficient. To solve these difficulties, nonlinear model predictive control (NMPC) attracted increasing attention over the past decade (Qin *et al.*, 2003, Cannon, 2004). Nowadays, the research on NMPC mainly focuses on its theoretical characters, such as stability, robustness and so on, while the computational method of NMPC is ignored in some extent. The fact mentioned above is one of the most serious reasons that obstruct the practical implementations of NMPC.

Analyzing the computational problem of NMPC, the direct incorporation of a nonlinear process into the linear MPC formulation structure may result in a non-convex nonlinear programming problem, which needs to be solved under strict sampling time constraints and has been proved as an NP-hard problem (Zheng, 1997). In general, since there is no accurate analytical solution to most kinds of nonlinear programming problem, we usually have to use numerical methods such as Sequential Quadric Programming (SQP) (Ferreau *et al.*, 2006) or Genetic Algorithm (GA) (Yuzgec *et al.*, 2006). Moreover, the computational load of NMPC using numerical methods is also much heavier than that of linear MPC, and it would even increase exponentially when the predictive horizon length increases. All of these facts lead us to develop a novel NMPC with analytical solution and little computational load in this chapter.

Since affine nonlinear system can represents a lot of practical plants in industry control, including the water-tank system that we used to carry out the simulations and experiments, it has been chosen for propose our novel NMPC algorithm. Follow the steps of research work, the chapter is arranged as follows:

In Section 2, analytical one-step NMPC for affine nonlinear system will be introduced at first, then, after description of the control problem of a water-tank system, simulations will be carried out to verify the result of theoretical research. Error analysis and feedback compensation will be discussed with theoretical analysis, simulations and experiment at last.

Then, in Section 3, by substituting reference trajectory for predicted state with stair-like control strategy, and using sequential one-step predictions instead of the multi-step prediction, the analytical multi-step NMPC for affine nonlinear system will be proposed. Simulative and experimental control results will also indicate the efficiency of it. The feedback compensation mentioned in Section 2 is also used to guarantee the robustness to model mismatch.

Conclusion and further research direction will be given at last in Section 4.

2. ONE-STEP NMPC FOR AFFINE SYSTEM

2.1. Description of Nmpc for affine system

Consider a time-invariant, discrete, affine nonlinear system with integer k representing the current discrete time event:

$$x_{k+1} = f(x_k) + g(x_k) \times u_k + \xi_k \tag{1a}$$

$$\text{s. t. } x_k \in X \subseteq R^n \tag{1b}$$

$$u_k \in U \subseteq R^m \tag{1c}$$

$$\xi_k \in R^n \tag{1d}$$

In the above, u_k, x_k, ξ_k are input, state and disturbance of the system respectively, $f{:}R^n \rightarrow R^n$, $g{:}R^n \rightarrow R^{n \times m}$, are corresponding nonlinear mapping functions with proper dimension.

Assume $\hat{x}_{k+j|k}$ are predictive values of x_{k+j} at time k, $\Delta u_k = u_k - u_{k-1}$ and $\Delta \hat{u}_{k+j|k}$ are the solutions of future increment of u_{k+j} at time k, then the objective function J_k can be written as follow:

$$J_k = F(\hat{x}_{k+p|k}) + \sum_{j=0}^{p-1} G(\hat{x}_{k+j|k}, \Delta u_{k+j|k}) \tag{2}$$

The function F (.) and G (.,.) represent the terminal state penalty and the stage cost respectively, where p is the predictive horizon.

In general, Jk usually has a quadratic form. Assume $w_{k+j|k}$ is the reference value of x_{k+j} at time k which is called reference trajectory (the form of $w_{k+j|k}$ will be introduced with detail in Section 2.2 and 3.1 for one-step NMPC and multi-step NMPC respectively), semi-positive definite matrix Q and positive definite matrix R are weighting matrices, (2) now can be written as:

$$J_k = \sum_{j=1}^{p} \left\| \hat{x}_{k+j|k} - w_{k+j|k} \right\|_Q^2 + \sum_{j=0}^{p-1} \left\| \Delta u_{k+j|k} \right\|_R^2$$

(3)

Corresponding to (1) and (3), the NMPC for affine system at each sampling time now is formulated as the minimization of Jk, by choosing the increments sequence of future control input $[\Delta u_{k|k} \quad \Delta u_{k+1|k} \quad \cdots \quad \Delta u_{k+p-1|k}]$, under constraints (1b) and (1c).

By the way, for simplicity, In (3), part of Jk is about the system state x_k, if the output of the system $y_k = C x_k$, which is a linear combination of the state (C is a linear matrix), we can rewrite (3) as follow to make an objective function Jk about system output:

$$J_k = \sum_{j=1}^{p} \left\| C\hat{x}_{k+j|k} - w_{k+j|k} \right\|_Q^2 + \sum_{j=0}^{p-1} \left\| \Delta u_{k+j|k} \right\|_R^2 = \sum_{j=1}^{p} \left\| \hat{y}_{k+j|k} - w_{k+j|k} \right\|_Q^2 + \sum_{j=0}^{p-1} \left\| \Delta u_{k+j|k} \right\|_R^2$$

(4)

And sometimes, $\Delta u_{k+j|k}$ in J_k in Jk could also be changed as $u_{k+j|k}$ to meet the need of practical control problems.

2.2. One-step Nmpc for affine system

Except for some special model, such as Hammerstein model, analytic solution of multi-step NMPC could not be obtained for most nonlinear systems, including the NMPC for affine system mentioned above in Section 2.1. But if the analytic inverse of system function exists (could be either state-space model or input-state model), the one-step NMPC always has the analytic solution. So all the research in this chapter is not only suitable for affine nonlinear system, but also suitable for other nonlinear systems, that have analytic inverse system function.

Consider system described by (1a-1d) again, the one-step prediction can be deduced directly as follow with only one unknown data $\Delta u_{k|k} = u_{k|k} - u_{k-1}$ at time k:

$$\hat{x}_{k+1|k} = f(x_k) + g(x_k) \cdot u_{k|k} = f(x_k) + g(x_k) \cdot u_{k-1} + g(x_k) \cdot \Delta u_{k|k} = \hat{x}_{k+1|k}^1 + g(x_k) \cdot \Delta u_{k|k}$$ (5)

In (5), $\hat{x}_{k+1|k}^1$ means the part which contains only known data (x_k and u_{k-1}) at time k, and $g(x_k) \cdot \Delta u_{k|k}$ is the unknown part of predictive state $\hat{x}_{k+1|k}$.

If there is no model mismatch, the predictive error of (5) will be $\tilde{x}_{k+1|k} = x_{k+1} - \hat{x}_{k+1|k} = \xi_{k+1}$.

Especially, if ξk is a stationary stochastic noise with zero mean and variance $E[\xi k]=\delta^2$, it is easy known that $E[\tilde{x}_{k+1|k}]=0$, and $E[(\tilde{x}_{k+1|k}-E[\tilde{x}_{k+1|k}])^T \cdot (\tilde{x}_{k+1|k}-E[\tilde{x}_{k+1|k}])]=n\delta^2$, in another word, both the mean and the variance of the predictive error have a minimum value, so the prediction is an optimal prediction here in (5).

Then if the setpoint is xsp, and to soften the future state curve, the expected state value at time k+1 is chosen as $w_{k+1|k} = \alpha x_k + (1-\alpha)x_{sp}$, where $\alpha \in [0,1)$ is called soften factor, thus the objective function of one-step NMPC can be written as follow:

$$J_k = \left\|\hat{x}_{k+1|k} - w_{k+1|k}\right\|_Q^2 + \left\|\Delta u_{k|k}\right\|_R^2$$

(6)

To minimize Jk without constraints (1b) and (1c), we just need to have $\frac{\partial J_k}{\partial \Delta u_{k|k}} = 0$ and $\frac{\partial^2 J_k}{\partial \Delta u_{k|k}^2} > 0$, then:

$$\Delta u_{k|k} = -(g(x_k)^T \cdot Q \cdot g(x_k) + R)^{-1} \cdot (g(x_k) \cdot Q \cdot (\hat{x}_{k+1|k}^1 - w_{k+1|k}))$$

(7)

Mark $H = g(x_k)^T \cdot Q \cdot g(x_k) + R$ and $F = g(x_k) \cdot Q \cdot (w_{k+1|k} - \hat{x}_{k+1|k}^1)$, so the increment of instant future input is:

$$\Delta u_{k|k} = -H^{-1}F$$

(8)

But in practical control problem, limitations on input and output always exist, so the result of (8) is usually not efficient. To satisfy the constraints, we can just put logical limitation on amplitudes of uk and xk, or some classical methods such as Lagrange method could be used. For simplicity, we only discuss about the Lagrange method here in this chapter.

First, suppose every constraint in (1b) and (1c) could be rewritten in the form as $a_i^T \Delta u_{k|k} \le b_i$, $i = 1,2,\cdots q$, then the matrix form of all constraints is:

$$A\Delta u_{k|k} \le B$$

(9)

In which, $A = \begin{bmatrix} a_1^T & a_2^T & \cdots & a_q^T \end{bmatrix}^T$ $B = \begin{bmatrix} b_1 & b_2 & \cdots & b_q \end{bmatrix}^T$.

Choose Lagrange function as $L_k(\lambda_i) = J_k + \lambda_i^T(a_i^T \Delta u_{k|k} - b_i)$, $i = 1,2,\cdots q$, let $\frac{\partial L}{\partial \Delta u_{k|k}} = H\Delta u_{k|k} + F + a_i\lambda_i = 0$ and $\frac{\partial L}{\partial \lambda_i} = a_i^T \Delta u_{k|k} - b_i = 0$, then:

$$\Delta u_{k|k} = -H^{-1}(F + a_i \lambda_i)$$

(10a)

$$\lambda_i = -\frac{a_i^T H^{-1} F + b_i}{a_i^T H^{-1} a_i}$$

(10b)

If $\lambda_i \leq 0$ in (10b), means that the corresponding constraint has no effect on $\Delta u_{k|k}$, we can choose $\lambda_i = 0$, but if $\lambda_i > 0$ in (10b), the corresponding constraint has effect on Δu_k indeed, so we must choose $\lambda_i = \bar{\lambda}_i$, finally, the solution of one-step NMPC with constraints could be:

$$\Delta u_{k|k} = -H^{-1}(F + A^T \bar{\Lambda})$$

(11)

In which, $\bar{\Lambda} = \begin{bmatrix} \bar{\lambda}_1 & \bar{\lambda}_2 & \cdots & \bar{\lambda}_q \end{bmatrix}^T$

2.3. Control problem of the water-tank system

Our plant of simulatiŏns and experiments in this chapter is a water-tank control system as that in Fig. 1. and Fig. 2. (We just used one water-tank of this three-tank system). Its affine nonlinear model is achieved by mechanism modeling (Chen *et al.*, 2006), in which the variables are normalized, and the sample time is 1 second here:

$$x_{k+1} = x_k - 0.2021\sqrt{x_k} + 0.01923 u_k$$

(12a)

$$\text{s. t. } x_k \in [0\%, 100\%]$$

(12b)

$$u_k \in [0\%, 100\%]$$

(12c)

In (12), x_k is the height of water in the tank, and u_k is the velocity of water flow into the tank, from pump P_1 and valve V_1, while valve V_2 is always open. In the control problem of the water-tank, for convenience, we choose the system state as the output, that means $y_k = x_k$, and the system functions are $f(x_k) = x_k - 0.2021\sqrt{x_k}$ and $g(x_k) = 0.01923$.

Figure 1. Photo of the water-tank system

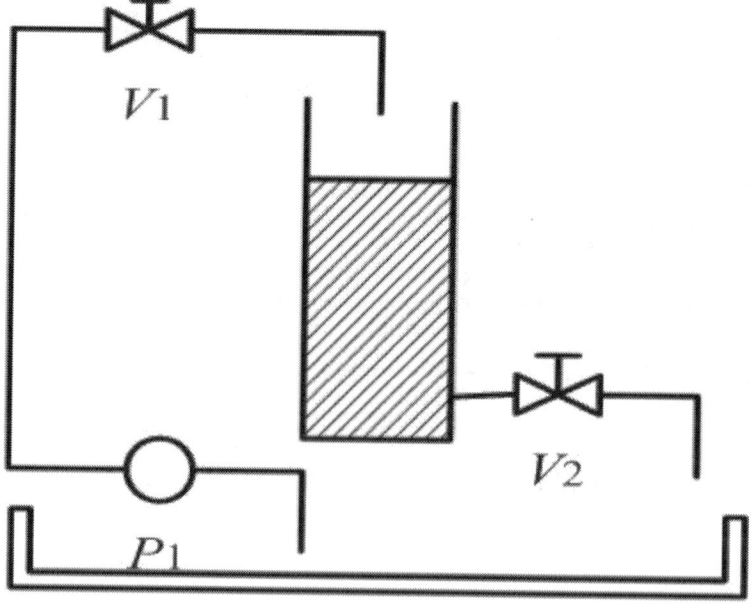

Figure 2. Structure of the water-tank system

To change the height of the water level, we can change the velocity of input flow, by adjusting control current of valve V_1, and the normalized relation between the control current and the velocity u_k is shown in Fig. 3.

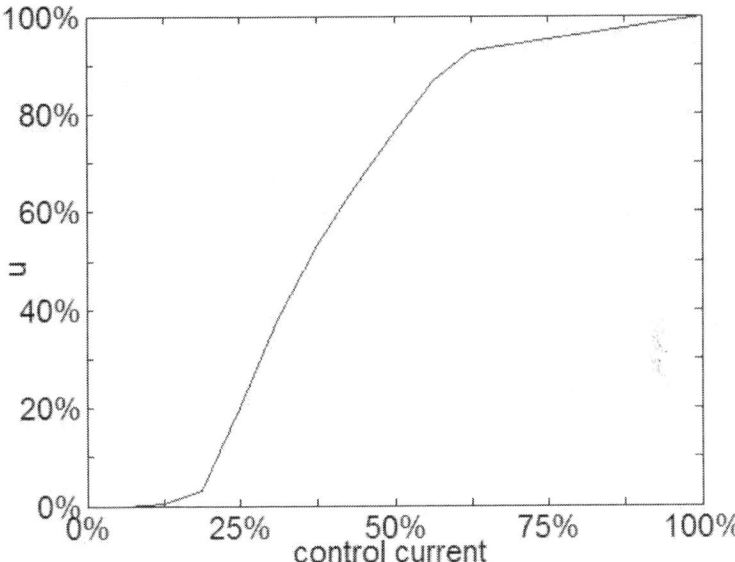

Figure 3. The relation between control current and input u_k

2.4 One-step NMPC of the water-tank system and its feedback compensation

Choose objective function $J_k = (\hat{x}_{k+1|k} - w_{k+1|k})^2 + 0.001\Delta u_{k|k}^2$, $x_{sp} = 30\%$ and soften factor $\alpha=0.95\alpha=0.95$, to carry out all the simulations and the experiment in this section. (except for part of Table 1., where we choose $\alpha=0.975\alpha=0.975$)

Suppose there is no model mismatch, the simulative control result of one-step NMPC for water-tank system is obtained as Fig. 4. and it is surely meet the control objective.

To imitate the model mismatch, we change the simulative model of the plant from $x_{k+1} = x_k - 0.2021\sqrt{x_k} + 0.01923u_k$ to $x_{k+1} = x_k - 110\% \times 0.2021\sqrt{x_k} + 90\% \times 0.01923u_k$, but still use $x_{k+1} = x_k - 0.2021\sqrt{x_k} + 0.01923u_k$ to be the predictive model in one-step NMPC, the result in Fig. 5. now indicates that there is obvious steady-state error.

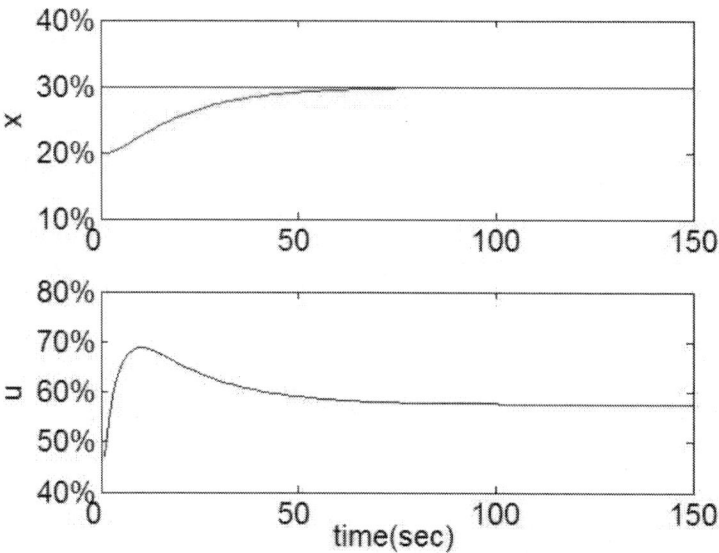

Figure 4. Simulation of one-step NMPC without model mismatch and feedback compensation

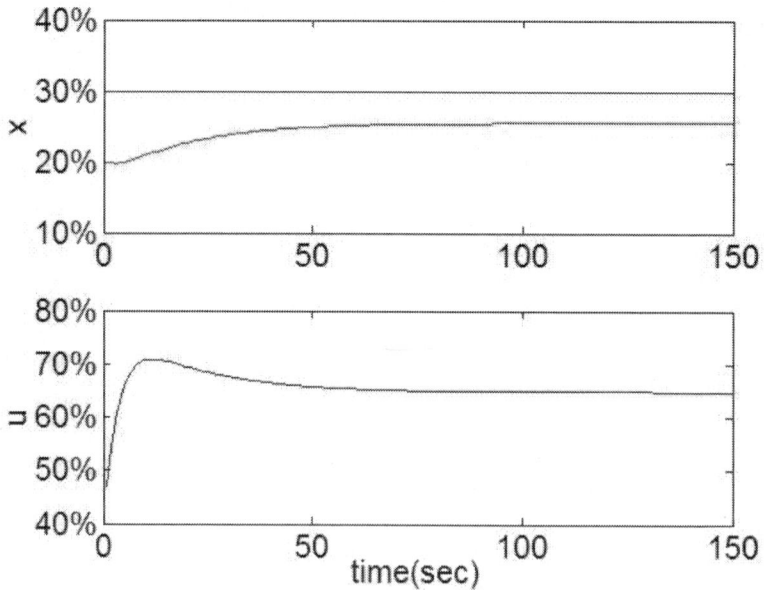

Figure 5. Simulation of one-step NMPC with model mismatch but without feedback compensation

Proposition 1: For affine nonlinear system $x_{k+1} = f'(x_k) + g'(x_k) \cdot u_k$, if the setpoint is x_{sp}, steady-state is u_s and x_s, and the predictive model is $x_{k+1} = f(x_k) + g(x_k) \cdot u_k$, without consideration of constraints, the steady-state error of one-step NMPC is $e = x_s - x_{sp} = \dfrac{(f'(x_s) - f(x_s)) + (g'(x_s) - g(x_s)) \cdot u_s}{1 - \alpha}$, in which α is the soften factor.

Proof: If the system is at the steady-state, then we have $x_{k-1} = x_k = x_s$ and $u_{k-1} = u_k = u_s$.

Since $u_{k-1} = u_k = u_s$, so $\Delta u_k = 0$, from (8), we know matrix F=0, or equally $(w_{k+1|k} - \hat{x}^1_{k+1|k}) = 0$.

Update the process of one-step NMPC at time k, we have:

$$w_{k+1|k} = \alpha x_k + (1 - \alpha) x_{sp} = \alpha x_s + (1 - \alpha) x_{sp}$$

(13)

$$\hat{x}^1_{k+1|k} = f(x_k) + g(x_k) \cdot u_{k-1} = f(x_s) + g(x_s) \cdot u_s$$

(14)

(13)-(14), and notice that $x_s = f'(x_s) + g'(x_s) \cdot u_s$ for steady-state, we get:

$$0 = \alpha x_s + (1 - \alpha) x_{sp} - f(x_s) - g(x_s) \cdot u_s = x_s - f(x_s) - g(x_s) \cdot u_s + (1 - \alpha) x_{sp} - (1 - \alpha) x_s$$

$$= (f'(x_s) - f(x_s)) + (g'(x_s) - g(x_s)) \cdot u_s + (1 - \alpha) \cdot (x_{sp} - x_s)$$

So:

$$e = x_s - x_{sp} = \frac{(f'(x_s) - f(x_s)) + (g'(x_s) - g(x_s)) \cdot u_s}{1 - \alpha}$$

(15)

Proof end.

Because the soften factor $\alpha \in [0,1)$, thus $1 - \alpha \neq 0$ always holds, the necessary condition for $e = 0$ is $(f'(x_s) - f(x_s)) + (g'(x_s) - g(x_s)) \cdot u_s = 0$. When there is model mismatch, there will be steady-state error, while this error is independent of weight matrix Q and dependent of the soften factor α. For corresponding discussion on steady-state error of one-step NMPC with constraints, the only difference is (11) will take the place of (8) in the proof.

Table 1. is the comparison on $e = x_s - x_{sp}$ between simulation and theoretical analysis, and they have the same result. (simulative model $x_{k+1} = x_k - 110\% \times 0.2021\sqrt{x_k} + 90\% \times 0.01923u_k$, predictive model $x_{k+1} = x_k - 0.2021\sqrt{x_k} + 0.01923u_k$)

Table 1. Comparison on $e = x_s - x_{sp}$ between simulation and theoretical analysis

α	Q	$e = x_s - x_{sp}$ Simulation(%)	$e = x_s - x_{sp}$ Value of (15)(%)
0.975	0	-8.3489	-8.3489
	0.001	-8.3489	-8.3489
	0.01	-8.3489	-8.3489
0.95	0	-4.5279	-4.5279
	0.001	-4.5279	-4.5279
	0.01	-4.5279	-4.5279

From (15) we know, we cannot eliminate this steady-state error by adjusting α, so feedback compensation could be used here, mark the predictive error e_k at time k as follow:

$$e_k = x_k - \hat{x}_{k|k-1} = x_k - (\hat{x}^1_{k|k-1} + g(x_{k-1}) \cdot \Delta u_{k-1}) \quad (16)$$

In which, x_k is obtained by system feedback at time k, and $\hat{x}_{k|k-1}$ is the predictive value of x_k at time k-1.

Then add k e to the predictive value of k 1 x + at time k directly, so (5) is rewritten as follow:

$$\hat{x}_{k+1|k} = f(x_k) + g(x_k) \cdot u_{k-1} + g(x_k) \cdot \Delta u_{k|k} + e_k = \hat{x}^1_{k+1|k} + g(x_k) \cdot \Delta u_{k|k} + e_k \quad (17)$$

Use this new predictive value to carry out one-step NMPC, the simulation result in Fig. 6. verify its robustness under model mismatch, since there is no steady-state error with this feedback compensation method.

The direct feedback compensation method above is easy to understand and carry out, but it is very sensitive to noise. Fig. 7. is the simulative result of it when there is noise add to the system state, we can see that the input vibrates so violently, that is not only harmful to the actuator in practical control system, but also harmful to system performance, because the actuator usually cannot always follow the input signal of this kind.

To develop the character of feedback compensation, simply, we can use the weighted average error $\bar{e}_k$ instead of single e_k in (17):

$$\hat{x}_{k+1|k} = \hat{x}^1_{k+1|k} + g(x_k) \cdot \Delta u_{k|k} + \sum_{i=1}^{s} h_i \cdot e_{k+1-i} = \hat{x}^1_{k+1|k} + g(x_k) \cdot \Delta u_{k|k} + \bar{e}_k, \quad \sum_{i=1}^{s} h_i = 1$$

(18)

Choose i=20, h_i=0.05, the simulative result is shown in Fig. 8. Compared with Fig. 7. it has almost the same control performance, but the input is much more smooth now. Using the same method and parameters, experiment has been done on the water-tank system, the result in Fig. 9. also verifies the efficiency of the proposed one-step NMPC for affine systems with feedback compensation.

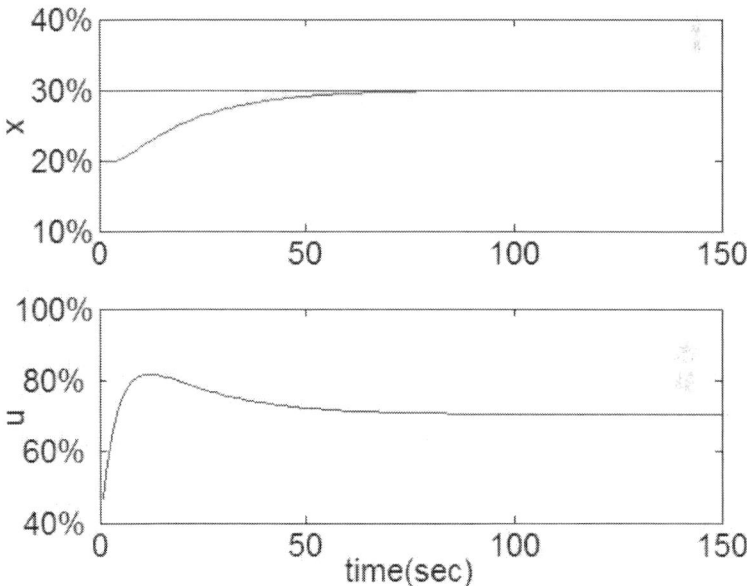

Figure 6. Simulation of one-step NMPC with model mismatch and direct feedback compensation

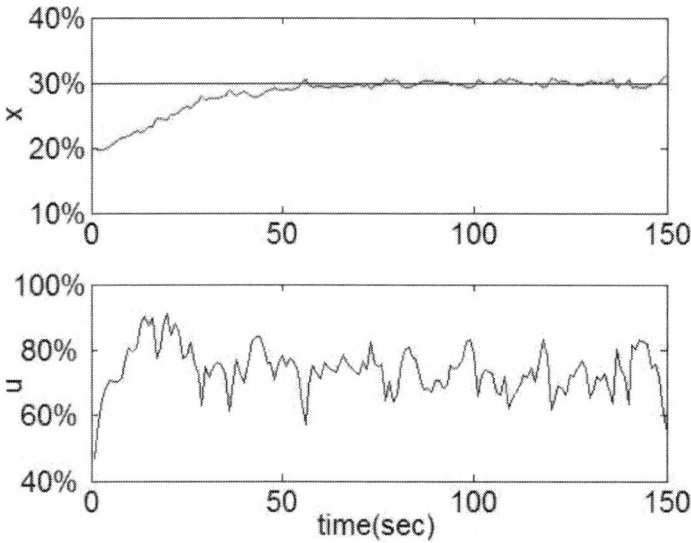

Figure 7. Simulation of one-step NMPC with model mismatch, noise and direct feedback compensation

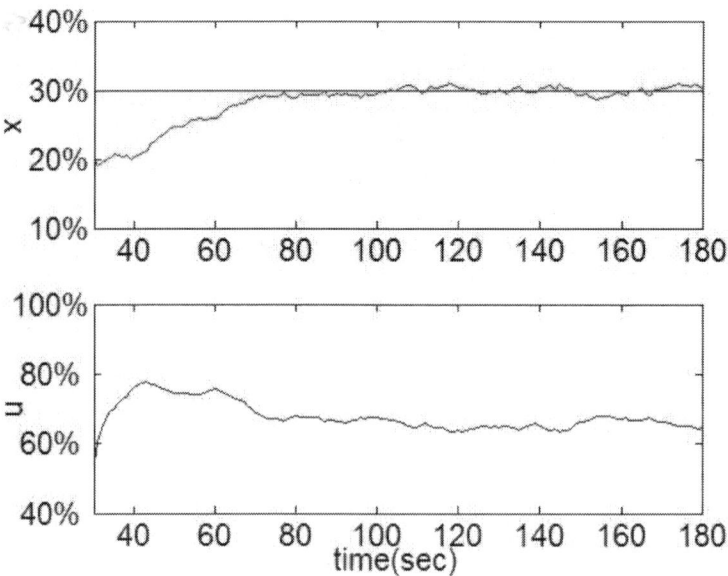

Figure 8. Simulation of one-step NMPC with model mismatch, noise and smoothed feedback compensation

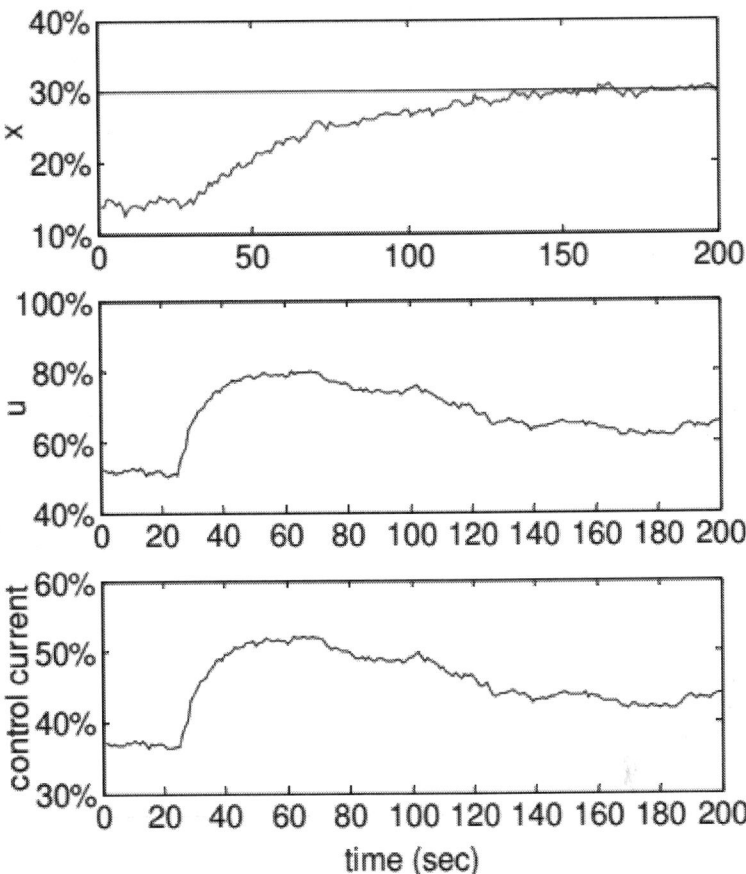

Figure 9. Experiment of one-step NMPC with setpoint $x_{sp}=30\%$

3. EFFICIENT MULTI-STEP NMPC FOR AFFINE SYSTEM

Since reference trajectory and stair-like control strategy will be used to establish efficient multi-step NMPC for affine system in this chapter, we will introduce them in Section 3.1 and 3.2 at first, and then, the multi-step NMPC algorithm will be discussed with theoretical research, simulations and experiments.

3.1. Reference trajectory for future state

In process control, the state usually meets the objective in the form of setpoint along a softer trajectory, rather than reach the setpoint immediately in only one sample time. This may because of the limit on control input, but a softer change of state is often more beneficial to actuators, even the whole process in practice. This trajectory, usually called reference trajectory, often can be defined as a first order exponential curve:

In which, x_{sp} still denotes the setpoint, $\alpha \in [0,1)$ is the soften factor, and the initial value of the trajectory is $w_{k|k} = x_k$. The value of α determines the speed of dynamic response and the curvature of the trajectory, the larger it is, the softer the curve is. Fig. 10. shows different trajectory with different α. Generally speaking, suitable α could be chosen based on the expected setting time in different practical cases.

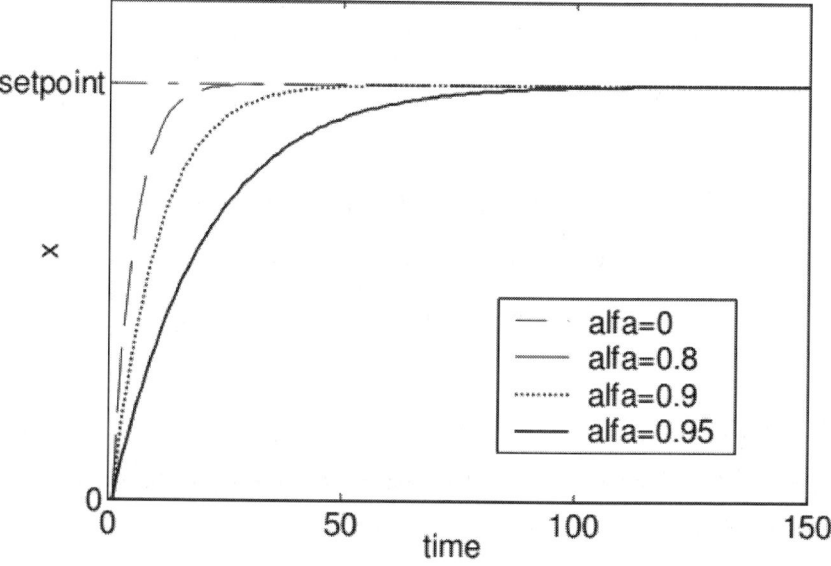

Figure 10. Reference trajectory with different soften factor α

3.2. Stair-like control strategy

To lighten the computational load of nonlinear optimization, which is one of the biggest obstacles in NMPC's application, stair-like control strategy is introduced here. Suppose the first unknown control input's increment $\Delta u_k = u_k - u_{k-1} = \Delta$, and the stair coefficient β is a positive real number, then the future control input's increment can be decided by the following expression:

$$\Delta u_{k+j} = \beta \cdot \Delta u_{k+j-1} = \beta^j \cdot \Delta u_k = \beta^j \cdot \Delta, j = 1, 2, \cdots, p-1$$

(20)

Instead of the full future sequence of control input's increment: $[\Delta u_k \quad \Delta u_{k+1} \quad \cdots \quad \Delta u_{k+p-1}]$, which has p independent variables. Using this strategy, in multi-step NMPC, it now need only compute Δu_k. The computational load now is independent of the length of predictive horizon, which is very convenient for us to choose long predictive horizon in NMPC to obtain a better control performance (Zheng *et al.*, 2007).

Since the dynamic optimization process will be repeated at every sample time, and only instant input $u_k = u_{k-1} + \Delta u_k$ will be carried out actually in NMPC, this strategy is efficient here. In the strategy, it supposes the future increase of control input will be in a same direction, which is the same as the experience in control practice of the human beings, and prevents the frequent oscillation of the input, which is very harmful to the actuators in real control plants. Fig. 11. shows the input sequences with different β.

3.3. Multi-step Nmpc for Affine system

The one-step NMPC in Section 2 is simple and fast, but it also has one fatal disadvantage. Its predictive horizon is only one step, while long predictive horizon is usually needed for better performance in MPC algorithms. One-step prediction may lead overshooting or other bad influence on system's behaviour. So we will try to establish a novel efficient multi-step NMPC based on proposed one-step NMPC in this section.

In this multi-step NMPC algorithm, the first step prediction is the same as (5), then follows the prediction of $\hat{x}_{k+1|k}$ in (5), the one-step prediction of $\hat{x}_{k+j|k}, j = 2, 3, \cdots, p$ could be obtained directly:

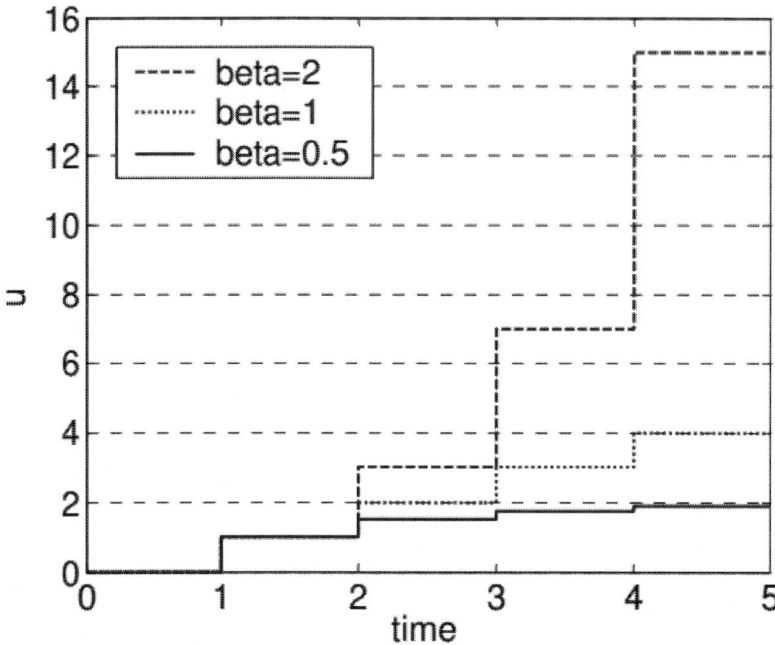

Figure 11. Stair-like control strategy

$$\hat{x}_{k+j|k} = f(\hat{x}_{k+j-1|k}) + g(\hat{x}_{k+j-1|k}) \cdot u_{k+j-1|k}$$
(21)

Since $\hat{x}_{k+j-1|k}$ already contains nonlinear function of former data, one may not obtain the analytic solution of (21) for prediction more than one step. Take the situation of j=2 for example:

$$\hat{x}_{k+2|k} = f(\hat{x}_{k+1|k}) + g(\hat{x}_{k+1|k}) \cdot u_{k+1|k} = f(f(x_k) + g(x_k) \cdot u_{k|k}) + g(f(x_k) + g(x_k) \cdot u_{k|k}) \cdot u_{k+1|k}$$
(22)

For most nonlinear f(.) and g(.), the embedding form above makes it impossible to get an analytic solution of u_{k+1} and further future input. So, using reference trajectory, we modified the one-step predictions when $j \geq 2$ as follow:

$$\hat{x}_{k+j|k} = f(w_{k+j-1|k}) + g(w_{k+j-1|k}) \cdot u_{k+j-1} = f(w_{k+j-1|k}) + g(w_{k+j-1|k}) \cdot (u_{k-1} + \sum_{i=0}^{j-1} \Delta u_{k+i|k})$$
(23)

Using the stair-like control strategy, mark $\Delta u_{k|k} = \Delta$, (23) can be transformed as:

$$\hat{x}_{k+j|k} = f(w_{k+j-1|k}) + g(w_{k+j-1|k}) \cdot \left(u_{k-1} + \sum_{i=0}^{j-1}\beta^i\Delta\right)$$

$$= f(w_{k+j-1|k}) + g(w_{k+j-1|k}) \cdot u_{k-1} + g(w_{k+j-1|k}) \cdot \sum_{i=0}^{j-1}\beta^i\Delta$$

$$= \hat{x}^1_{k+j|k} + g(w_{k+j-1|k}) \cdot \sum_{i=0}^{j-1}\beta^i\Delta$$

$$(24)$$

Here, $\hat{x}^1_{k+j|k}$ contains only the known data at time k, while the other part is made up by the increment of future input, thus the unknown data are separated linearly by (24), so the analytic solution of Δ can be achieved.

For $j = 1, 2, \cdots, p$, write the predictions in the form of matrix:

$$\hat{X}_k = \begin{bmatrix} \hat{x}_{k+1|k} \\ \hat{x}_{k+2|k} \\ \vdots \\ \hat{x}_{k+p|k} \end{bmatrix} \quad X^1_k = \begin{bmatrix} \hat{x}^1_{k+1|k} \\ \hat{x}^1_{k+2|k} \\ \vdots \\ \hat{x}^1_{k+p|k} \end{bmatrix} \quad W_k = \begin{bmatrix} w_{k+1|k} \\ w_{k+2|k} \\ \vdots \\ w_{k+p|k} \end{bmatrix} \quad \Delta U_k = \begin{bmatrix} \Delta u_{k|k} \\ \Delta u_{k+1|k} \\ \vdots \\ \Delta u_{k+p-1|k} \end{bmatrix} = \begin{bmatrix} \Delta \\ \beta\Delta \\ \vdots \\ \beta^{p-1}\Delta \end{bmatrix}$$

$$S_k = \begin{bmatrix} g(w_{k|k}) = g(x_k) & 0 & \cdots & 0 \\ g(w_{k+1|k}) & g(w_{k+1|k}) & \cdots & 0 \\ \vdots & \vdots & \ddots & \vdots \\ g(w_{k+p-1|k}) & g(w_{k+p-1|k}) & \cdots & g(w_{k+p-1|k}) \end{bmatrix} = \begin{bmatrix} s_1 & 0 & \cdots & 0 \\ s_2 & s_2 & \cdots & 0 \\ \vdots & \vdots & \ddots & \vdots \\ s_p & s_p & \cdots & s_p \end{bmatrix}$$

$$S_k \cdot \Delta U_k = \begin{bmatrix} s_1 \\ s_2(1+\beta) \\ \vdots \\ s_p(1+\beta+\cdots+\beta^{p-1}) \end{bmatrix} \quad \Delta = \bar{S}_k \cdot \Delta$$

$$(25)$$

Thus $\hat{X}_k = X^1_k + S_k \cdot \Delta U_k = X^1_k + \bar{S}_k \cdot \Delta$, for minimization of traditional quadric objective function

$\min_{\Delta} J_k = \min_{\Delta}[(\hat{X}_k - W_k)^T Q(\hat{X}_k - W_k) + \Delta U_k^T R \Delta U_k]$, where semi-positive definite matrix Q and positive definite matrix R are weighting matrices,

by $\frac{\partial J_k}{\partial \Delta} = 0$ and $\frac{\partial^2 J_k}{\partial \Delta^2} > 0$, the control solution of multi-step prediction is then obtained. Especially for single input problem, with objective function $\min_{\Delta} J_k = \min_{\Delta}[(\hat{X}_k - W_k)^T(\hat{X}_k - W_k) + r\Delta U_k^T \Delta U_k]$, it is easily denoted as follow:

$$\Delta = \frac{\bar{S}_k^T(W_k - X_k^1)}{\bar{S}_k^T \bar{S}_k + r(1 + \beta^2 + \cdots + \beta^{2(p-1)})}$$

(26)

At last, the instant input $u_{k|k} = u_{k-1} + \Delta$ can be carried out actually. As mentioned in Section 2, and if the model mismatch can be seen as time-invariant in p sample time (usually satisfied in the case of steady state in practice), to maintain the robustness, e_k or $\bar{e}_k$ can be also added to every prediction as mentioned in (17) and (18):

$$\hat{x}_{k+1|k} = f(x_k) + g(x_k) \cdot u_{k|k} + \bar{e}_k$$

$$\hat{x}_{k+j|k} = f(w_{k+j-1|k}) + g(w_{k+j-1|k}) \cdot u_{k+j-1|k} + \bar{e}_k, j = 2, 3, \cdots, p$$

(27)

Though there are approximate predictions in the novel NMPC which may take in some inaccuracy, the feedback compensation mentioned above and the new optimization process at every sample time will eliminate the error before its accumulation, to keep the efficiency of the algorithm. The constraints also could be handled by methods mentioned Section 2 or by other numerical optimizing algorithm, thus we would not discuss about it here again.

3.4. Multi-step NMPC of the water-tank system

Choose $\alpha=0.975$, $\beta=0.5$, $r=0.005$, $x_{sp}=60\%\text{or}30\%$ and predictive horizon $p=10$ to carry out simulations. Still use the different plant model and predictive model as that of Fig. 5. and Fig. 6. to imitate the model mismatch, the result in Fig. 12. and Fig. 13. shows the efficiency and robustness of this efficient multi-objective NMPC.

Choose $\alpha=0.975$, $\beta=0.5$, $r=0.005$, $x_{sp}=60\%$ to carry out experiments. Comparing control result between one-step NMPC and multi-step NMPC in Fig. 14. and Fig. 15., we can see the obvious developments on both input and output of the water-tank system when longer predictive horizon is used. It also verifies the efficiency of proposed novel multi-step NMPC algorithm. At last, Fig. 16. is the satisfactory performance of the efficient multi-step NMPC under disturbance (we open an additional outlet valve of the tank for 20 seconds).

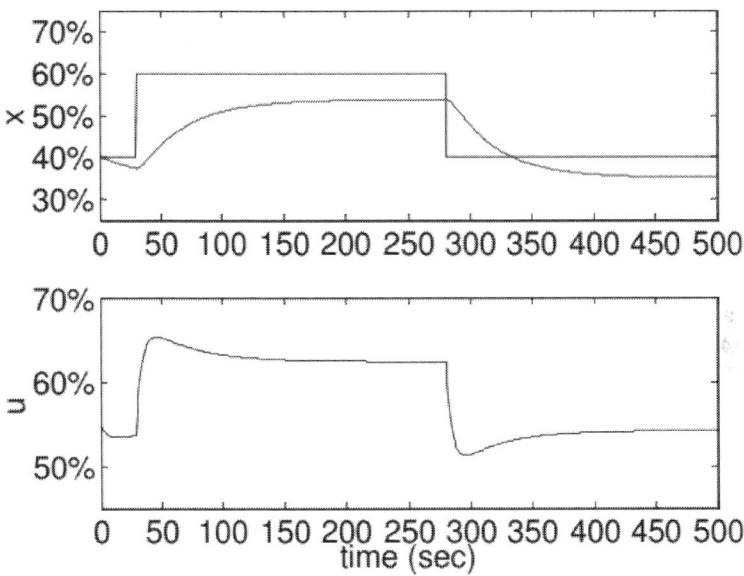

Figure 12. Simulation of multi-step NMPC with model mismatch but without feedback compensation

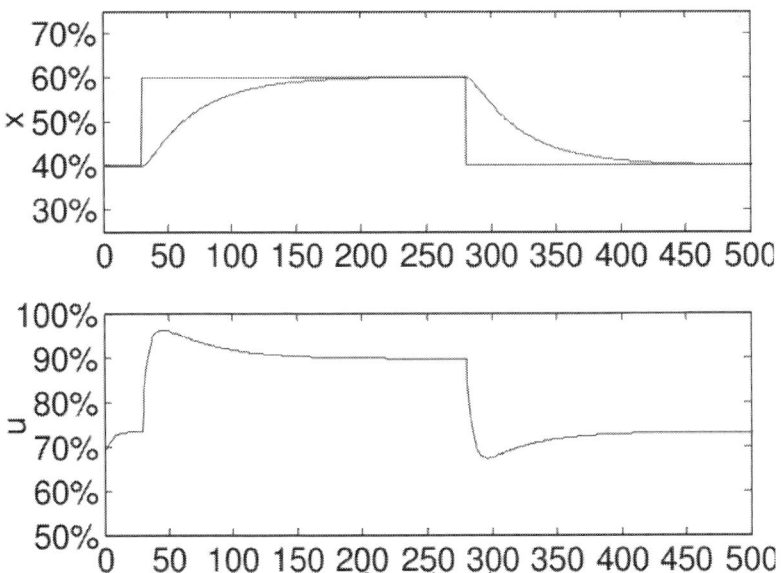

Figure 13. Simulation of multi-step NMPC with model mismatch and feedback compensation

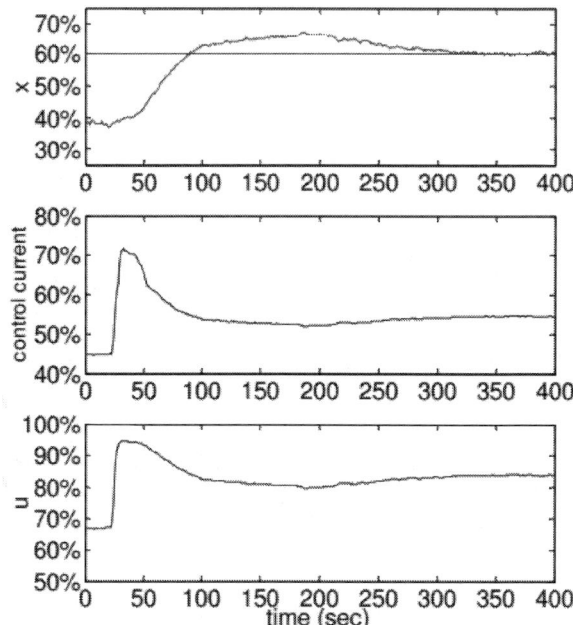

Figure 14. Experiment of one-step NMPC with setpoint x_{sp}=60%(Overshooting exists when setpoint is higher than Fig. 9.)

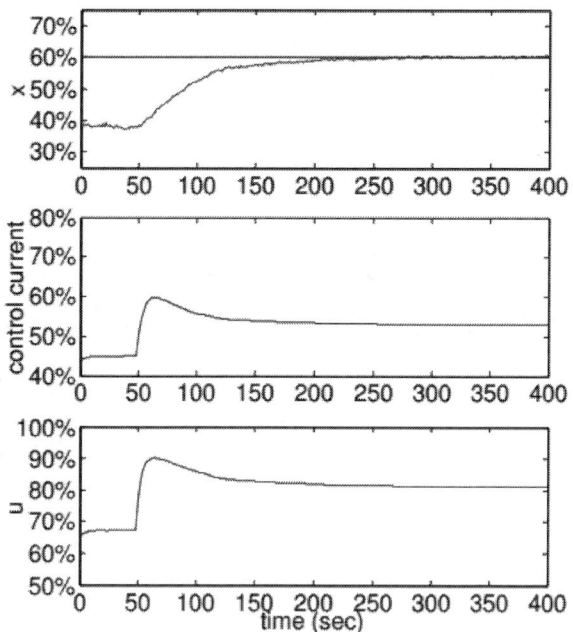

Figure 15. Experiment of one-step NMPC with setpoint x_{sp}=60% (p=10 and and No overshooting)

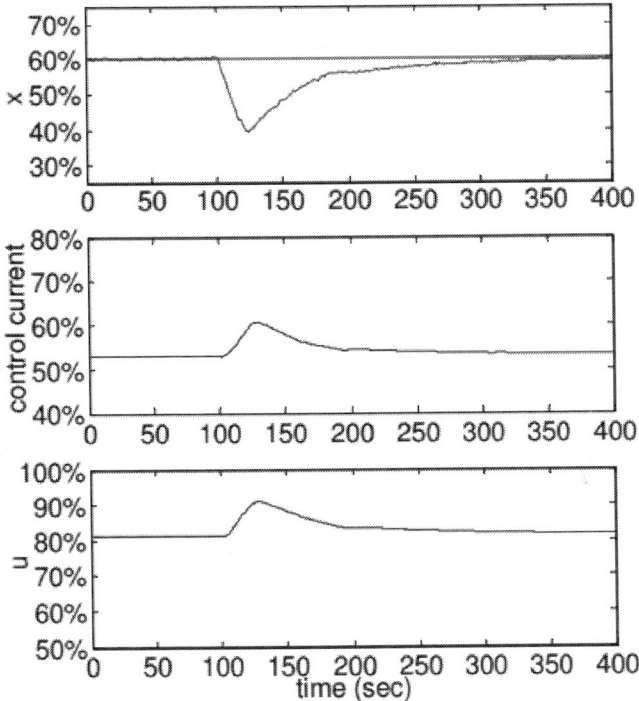

Figure 16. Experiment of one-step NMPC under disturbance

4. CONCLUSION

Using a series of approximate one-step predictions instead of the traditional multi-step prediction, the proposed multi-step NMPC leaded to an analytic result for nonlinear control of affine system. The use of stair-like control strategy caused a very little computational load and the feedback compensation brought robustness of model mismatch to it.

The simulations and experiments verify the practicability and efficiency of this multi-step NMPC for affine system, while the theoretical stability and other analysis will be the future work with considerable value.

ACKNOWLEDGEMENTS

This work is supported by National Natural Science Foundation of China (Youth Foundation, No. 61004082) and Special Foundation for Ph. D. of Hefei University of Technology (No. 2010HGBZ0616, from the Fundamental Research Funds for the Central Universities).

REFERENCES

1. M. Cannon, 2004 Efficient nonlinear model predictive control algorithms. Annual Reviews in Control, 28 2 January, 2004, 229 237 , 1367-5788
2. W. Chen, G. Wu, 2006 Modeling of nonlinear two-tank system and model predictive control. Journal of System Simulation, 18 8 August, 2006, 2078 2081 , 0100-4731X
3. H. J. Ferreau, G. Lorini, M. Dieh,l, 2006 Fast nonlinear model predictive control of gasoline engines. Proceedings of the 2006 IEEE International Conference on Control Applications, 0-78039-795-9 2754 2759, Munich, Germany, October, 2006, IEEE Publishing, Piscataway, USA.
4. S. J. Qin, T. A. Badgwell, 2003 A survey of industrial model predictive control technology. Control Engineering Practice, 11 7 July, 2003, 733 764 , 0967-0661
5. U. . Yuzgec, Y. Becerikl, M. Turker, 2006 Nonlinear predictive control of a drying process using genetic algorithms. ISA Transactions, 45 4 October, 2006, 589 602, 0019-0578
6. Zheng, 1997 A Computationally Efficient Nonlinear MPC Algorithm. Proceedings of the American Control Conference, 0-78033-832-4 1623 1627, Albuquerque, USA, June, 1997, IEEE Publishing, Piscataway, USA.
7. T. Zheng, G. Wu, D. F. He, 2007 Nonlinear Model Predictive Control Based on Lexicographic Multi-objective Genetic Algorithm. Proceedings of International Conference on Intelligent & Advanced Systems 2007, 1-42441-355-9 61 65, Kuala Lumpur, Malaysia, November, 2007, IEEE Publishing, Piscataway, USA.

CHAPTER 2
Application of Model Predictive Control to BESS for Microgrid Control

Thai-Thanh Nguyen, Hyeong-Jun Yoo and Hak-Man Kim*

Department of Electrical Engineering, Incheon National University, 12-1 Songdo-dong, Yeonsu-gu, Incheon 406-840, Korea

ABSTRACT

Battery energy storage systems (BESSs) have been widely used for microgrid control. Generally, BESS control systems are based on proportional-integral (PI) control techniques with the outer and inner control loops based on PI regulators. Recently, model predictive control (MPC) has attracted attention for application to future energy processing and control systems because it can easily deal with multivariable cases, system constraints, and nonlinearities. This study considers the application of MPC-based BESSs to microgrid control. Two types of MPC are presented in this study: MPC based on predictive power control (PPC) and MPC based on PI control in the outer and predictive current control (PCC) in the inner control loops. In particular, the effective application of MPC for microgrids with multiple BESSs should be considered because of the differences in their control performance. In this study, microgrids with two BESSs based on two MPC techniques are considered as an example. The control performance of the MPC used for the control microgrid is compared to that of the PI control. The proposed control strategy is investigated through simulations using MATLAB/Simulink software. The simulation results show that the response time, power and voltage ripples, and frequency spectrum could be improved significantly by using MPC.

Keywords: Microgrid; Model predictive control; Predictive power control; Battery energy storage system (BESS); Frequency control

1. INTRODUCTION

Microgrids are becoming popular in distribution systems because they can improve the power quality and reliability of power supplies and reduce the environmental impact. Microgrid operation can be classified into two modes: grid-connected and islanded modes. In general, microgrids are comprised of distributed energy resources (DERs) including renewable energy sources, distributed energy storage systems (ESSs), and local loads [1,2,3]. However, the use of renewable energy sources such as wind and solar power in microgrids causes power flow variations owing to uncertainties in their power outputs. These variations should be reduced to meet power-quality requirements [4,5]. This study focuses on handling the problems that are introduced by wind power.

To compensate for fluctuations in wind power, various ESSs have been implemented in microgrids. Short-term ESSs such as superconducting magnetic energy storage (SMES) systems [6], electrical double-layer capacitors (EDLCs) [7], and flywheel energy storage systems (FESSs) [8,9,10] as well as long-term ESSs such as battery energy storage systems (BESSs) [11,12] are applied to microgrid control. ESSs can also be used to control the power flow at point of common coupling in the grid-connected mode as well as to regulate the frequency and voltage of a microgrid in the islanded mode. Among these ESSs, BESSs have been implemented widely owing to their versatility, high energy density, and efficiency. Moreover, their cost has decreased whereas their performance and lifetime has increased [13].

In practice, BESSs with high performance such as smooth and fast dynamic response during charging and discharging are required for microgrid control. This performance depends on the control performance of the power electronic converter. Proportional-integral (PI) control is a practical and popular control technique for BESS control systems. However, PI control might show unsatisfactory results for nonlinear and discontinuous systems [10]. Meanwhile, model predictive control (MPC) is considered an attractive alternative to promote the performance of future energy processing and control systems [14]. Predictive strategies are based on the inherent discrete nature of a power converter. Owing to the finite number of switching states of a power converter, all possible states are considered for predicting the system behavior. Then, each prediction is used to evaluate a cost function. Consequently, the switching state with the minimum cost function is selected and applied to the converter [15]. One of the advantages of an MPC is the easy inclusion of constraints and nonlinearities. Therefore, MPC has been widely applied to drive applications [15,16,17,18] and power converters such as active front-end rectifiers [19], matrix converters [20], and multilevel converters [21]. Recently, it has been applied to a bidirectional AC-DC converter for use in BESSs [22,23,24].

Only a few literatures were found on the application of MPC to microgrid control. Most existing studies focused on MPC for a distributed generator in a microgrid with voltage and/or power control [25,26,27]. A modified MPC method for voltage control of a BESS in the islanded mode operation of a microgrid was presented in [27]. However, this study did not deal with frequency control in the islanded mode operation of a microgrid. MPC based on PI control in the outer control loop and predictive current control (PCC) in the inner control loop for BESS was presented in [28]. Coordinated predictive control of a wind/battery microgrid system was proposed to maintain the system voltage and frequency by adjusting the output power of BESS. PCC was used to control the current in the inner control loop, whereas PI regulators were used to regulate the voltage and power in the outer control loop. Owing to the use of PI regulators in the outer control loop, the dynamic response time under such MPC techniques was similar to that under PI control techniques with outer and inner control loops using PI regulators.

Another MPC technique is based on predictive power control (PPC), in which the power is predicted and controlled directly. This MPC technique could be applied to microgrid control because it affords advantages such as fast dynamic response for power control; however, studies have not yet explored the application of the PPC-based MPC technique to microgrid control. Furthermore, this MPC technique can only be used for power control. To overcome this problem, PI regulators can be used in an additional control loop to control the frequency and voltage. Therefore, this MPC technique uses PI regulators in the outer control loop and PPC in the inner control loop. It is similar to previous MPC techniques in which PI control is used in the outer control loop and PCC is used in the inner control loop. However, an MPC technique based on PI and PPC requires more computation time than does one based on PI and PCC, owing to the predicting powers in the inner PPC control loop. Therefore, in a microgrid with a single BESS, MPC based on PI and PCC is a suitable alternative for microgrid control. Another approach to overcome this limitation of the MPC control technique is to use a droop control scheme. Thus, a PPC-based MPC technique can be applied to microgrids consisting of multiple BESSs with different functionalities. This study deals with the effective application of an MPC technique to a microgrid with two BESSs as an example of multiple BESSs in a microgrid.

This study discusses the effective application of two MPC techniques to BESSs for microgrid control based on the characteristics of the MPC techniques as well as the functionalities of BESSs. One BESS is based on PI control in the outer and PCC in the inner control loops (PI (outer) + PCC (inner)); it is used for smoothing wind power fluctuations both in the grid-connected and the islanded modes. The other BESS is based on PPC (one loop); it controls the tie-line powers at the point of common coupling in the grid-connected mode and the frequency in the islanded mode. Additionally, to reduce the power losses of converters, the reduction of the switching frequency of the converter is considered an additional control variable in the MPC algorithm. The control performances of the two types of MPC techniques are compared to the PI control technique using PI regulators in the outer and inner control loops (PI (outer) + PI (inner)). The tuning of PI regulator parameters must be taken into account to effectively compare the control performance of MPC techniques to the PI control technique. Several tuning techniques have been used to select the PI regulator parameters. In this study, the tuning technique provided by MATLAB/Simulink software is used. The efficacy of the proposed control system is verified via simulations in the MATLAB/Simulink environment.

The remainder of this paper is organized as follows. Section 2 introduces the discrete-time model of the converter for prediction and MPC algorithms. Two types of MPC techniques are introduced in this section. Section 3 describes the microgrid system used to test the performance of the proposed control strategies. Section 4presents a comparison of the MPC and PI control techniques and the considerations for the effective application of MPC-based BESSs to microgrid control. Section 5 presents the simulation results for microgrid control in the grid-connected and islanded modes. The performances of the MPC techniques are compared to those of the PI control technique. Finally, Section 6 summarizes the main conclusions of this study.

2. MPC FOR BESS

2.1. Discrete-Time Model of Converter

The predicted variables of BESS are determined based on the discrete-time model of the converter. In this study, the BESS uses a two-level voltage source converter (VSC) converter, shown in Figure 1, connected to the three-phase AC power supply voltage v_g through filter inductance L and resistance R. The equations for each phase are given by Equations (1)–(3):

$$v_{aN} = L\frac{di_a}{dt} + Ri_a + v_{ga}$$

$$\tag{1}$$

$$v_{bN} = L\frac{di_b}{dt} + Ri_b + v_{gb} \tag{2}$$

$$v_{cN} = L\frac{di_c}{dt} + Ri_c + v_{gc}$$

$$\tag{3}$$

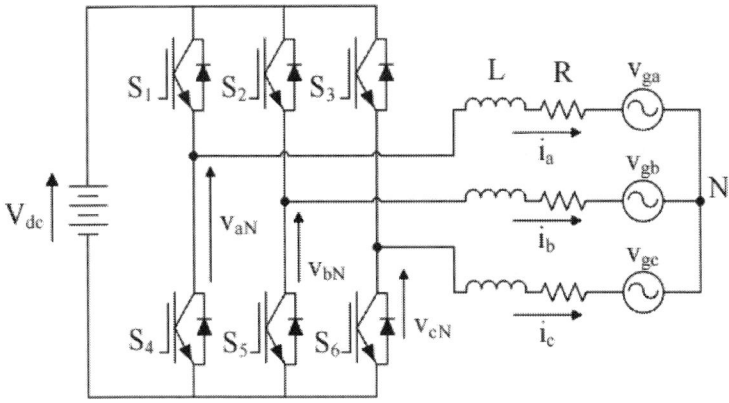

Figure 1. Configuration of BESS.

These equations can be represented by the space-vector equations given in Equation (4).

$$\frac{2}{3}(v_{aN} + av_{bN} + a^2 v_{cN}) = L\frac{d}{dt}\left(\frac{2}{3}(i_a + ai_b + a^2 i_c)\right) + R\left(\frac{2}{3}(i_a + ai_b + a^2 i_c)\right) +$$

$$\frac{2}{3}(v_{ga} + av_{gb} + a^2 v_{gc}) \qquad (4)$$

where a=ej2π/3.

Equation (4) can be simplified by considering the following definitions.

$$v = \frac{2}{3}(v_{aN} + av_{bN} + a^2 v_{cN}) \qquad (5)$$

$$i = \frac{2}{3}(i_a + ai_b + a^2 i_c) \qquad (6)$$

$$v_g = \frac{2}{3}(v_{ga} + av_{gb} + a^2 v_{gc}) \qquad (7)$$

The voltage v in Equation (5) is determined by the switching states of the converter and the DC link voltage (VDC), as given in Equation (8).

$$v = \frac{2}{3}V_{DC}(S_a + aS_b + a^2 S_c) \qquad (8)$$

Where the switching signals S_a, S_b, and S_c are defined as follows:

$$S_a = \begin{cases} 1 \text{ if } S_1 \text{ on and } S_4 \text{ off} \\ 0 \text{ if } S_1 \text{ off and } S_4 \text{ on} \end{cases} \tag{9}$$

$$S_b = \begin{cases} 1 \text{ if } S_2 \text{ on and } S_5 \text{ off} \\ 0 \text{ if } S_2 \text{ off and } S_5 \text{ on} \end{cases} \tag{10}$$

$$S_c = \begin{cases} 1 \text{ if } S_3 \text{ on and } S_6 \text{ off} \\ 0 \text{ if } S_3 \text{ off and } S_6 \text{ on} \end{cases} \tag{11}$$

The combination of S_a, S_b, and S_c creates eight switching states and eight voltage vectors, as shown in Table 1.

Table 1. Switching states and voltage vectors [29].

x	S_a	S_b	S_c	Voltage vectors v
1	0	0	0	$v_0 = 0$
2	1	0	0	$v_1 = \dfrac{2}{3} V_{dc}$
3	1	1	0	$v_2 = \dfrac{1}{3} V_{dc} + j \dfrac{\sqrt{3}}{3} V_{dc}$
4	0	1	0	$v_3 = -\dfrac{1}{3} V_{dc} + j \dfrac{\sqrt{3}}{3} V_{dc}$
5	0	1	1	$v_4 = -\dfrac{2}{3} V_{dc}$
6	0	0	1	$v_5 = -\dfrac{1}{3} V_{dc} - j \dfrac{\sqrt{3}}{3} V_{dc}$
7	0	1	1	$v_6 = \dfrac{1}{3} V_{dc} - j \dfrac{\sqrt{3}}{3} V_{dc}$
8	1	1	1	$v_7 = 0$

Substituting Equations (5)–(7) in Equation (4), we get

$$v = L\frac{di}{dt} + Ri + v_g \tag{12}$$

From Equation (12), the discrete-time model of the converter is determined by approximating the derivative load current di/dt in terms of a forward Euler approximation, as shown in Equation (13).

$$\frac{di}{dt} \approx \frac{i(k+1) - i(k)}{T_s} \tag{13}$$

By substituting Equation (13) in Equation (12), the future current at the sampling instant $k + 1$ is represented as

$$i^p(k+1) = \left(1 - \frac{RT_s}{L}\right)i(k) + \frac{T_s}{L}\left(v(k) - v_g(k)\right) \tag{14}$$

where $i(k)$ and $v_g(k)$ are the three-phase current and voltage of the BESS measured at sampling instant k, respectively; $v(k)$ is the voltage vector according to the eight switching states of the converter; and T_s is the sampling time.

Based on the measured voltage and current of BESS at sampling instant k, the variables at sampling instant $k + 1$ are predicted as given in Equation (14). For a small sampling time (T_s), the predicted grid voltage at sampling instant $k + 1$ can be assumed equal to the measured grid voltage at the kth sampling instant $(v_g(k + 1) = v_g(k))$ owing to the fundamental grid frequency [29]. As a result, the predicted instantaneous real and reactive powers can be expressed as follows:

$$P^p(k+1) = 1.5 Re\left\{\bar{i}^p(k+1)v_g^m(k)\right\} \tag{15}$$

$$Q^p(k+1) = 1.5 Im\left\{\bar{i}^p(k+1)v_g^m(k)\right\} \tag{16}$$

where $\overline{i}^{\,p}(k+1)$ is the complex conjugate of the predicted current vector $i^p(k+1)$.

Equations (14)–(16) show that the predictive current and power highly rely on system model, converter, and filter parameters. Any change in the model parameters can provide inaccuracy in the predictive variables. Reference [29] shown that the current or power ripple could be affected by the parameter variations, whereas the dynamic response was almost unchanged. In case of extreme variations in the model parameters, an online parameter estimation algorithm should be included in the MPC strategy [30,31]. However, MPC can effectively handle the small change in inductive filter parameters. The comparison between MPC with and without the online filter estimation was presented in [29,32]. The major errors were observed at low values of the filter parameters. In addition, only a small difference was observed at high values of the filter parameters. In this study, a high value of the filter parameter is chosen to avoid the major errors by filter parameter variations. Thus, the model parameters is assumed unchanged during simulation for the sake of simplicity.

2.2. Principle of MPC

MPC is based on the inherent discrete nature of a power converter, which has a finite number of switching states. All possibilities of variables (current or real/reactive powers) of the converter according to switching states can be predicted. The predicted variables are compared to the reference control signal, and the predicted variable that is closest to the reference control signal is chosen as shown in Figure 2. Then, the switching state related to this predicted variable is applied to control the converter.

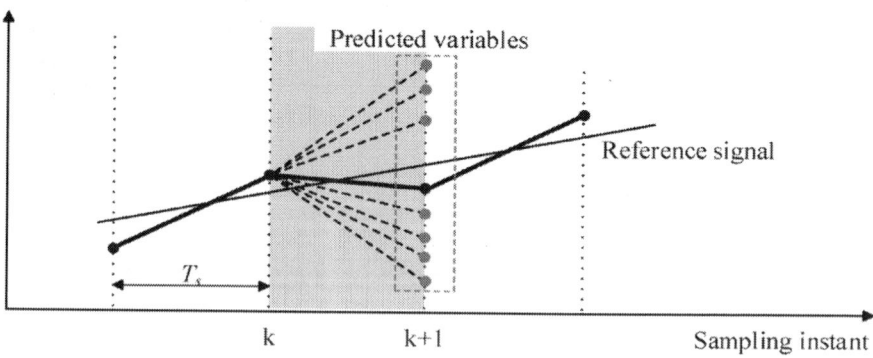

Figure 2. Principle of MPC.

Figure 3 shows two types of MPC techniques applied for BESSs: MPC based on PI control in the outer and PCC in the inner control loops (Figure 3a) and MPC based on PPC (Figure 3b). As shown in Figure 3a, PI control in the outer control loop is used to regulate the real/reactive powers as well as voltage of the microgrid. The reference current obtained by the outer control loop is used for the inner PCC control loop based on Equation (14). As shown in Figure 3b, in comparison, PPC based on Equations (15) and (16) can control real/reactive powers directly. To control the frequency of the microgrid, the frequency droop control scheme is suitable for a BESSs control system. However, conventional droop control can cause a steady-state error [33]. Thus, this study proposes an improved droop control scheme in which the steady-state error is removed by a new feedback signal through the PI regulator [9].

The objective of the MPC scheme is to minimize the error between the reference values and the measured values. This can be achieved by introducing a cost function gC for PCC and gS for PPC, as shown in the following equations.

$$g_C = \left| i_\alpha^*(k+1) - i_\alpha^p(k+1) \right|^2 + \left| i_\beta^*(k+1) - i_\beta^p(k+1) \right|^2 + \lambda_C \cdot n$$

$$(17)$$

$$g_S = |P^*(k+1) - P^p(k+1)|^2 + |Q^*(k+1) - Q^p(k+1)|^2 + \lambda_S \cdot n$$

$$(18)$$

where $i_\alpha^*(k+1)$ and $i_\beta^*(k+1)$ are the real and imaginary parts of the reference current, $i_\alpha^p(k+1)$ and $i_\beta^p(k+1)$ are the real and imaginary parts of the predicted current vectors $i^p(k+1)$ according to Equation (14), $P^*(k+1)$ and $Q^*(k+1)$ are the real and reactive reference powers, $P^p(k+1)$ and $Q^p(k+1)$ are the predicted real and reactive powers according to Equations (15) and (16), $\lambda_C \cdot n$ and $\lambda_S \cdot n$ represent the reduction of switching frequency of the converter where n is the number of switches that change when the switching states $S = (S_a, S_b, S_c)$ are applied, and λ_C and λ_S are the weighting factor for PCC and PPC, respectively.

The cost functions g_C and g_S have two terms with different goals. The primary goal is the current control in case of g_C or power control in case of g_S, which must be achieved to provide a proper system behavior. The secondary goal is the reduction of switching frequency ($\lambda_C \cdot n$ and $\lambda_S \cdot n$) in both cost functions. The importance of second term corresponds to the weighting factors λ_C and λ_S that can impose a trade-off with the primary control objective. The algorithm to adjust the weighting factors proposed in [29] is used in this study. Total harmonic distortion (THD) is used to estimate the trade-off between the primary and secondary goals.

The switching frequency of the converter depends on the change in the switching state, which can be only one or zero. Therefore, the number of switches that change from $S(k-1)$ to $S(k)$ is defined as given in Equation (19):

$$n = |S_a(k) - S_a(k-1)| + |S_b(k) - S_b(k-1)| + |S_c(k) - S_c(k-1)|$$

$$(19)$$

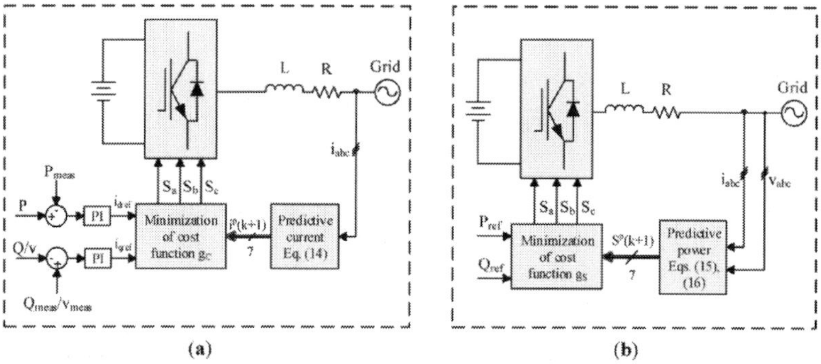

(a)   (b)

Figure 3. MPC block diagrams: (a) MPC based on PI control in the outer and PCC in the inner control loops; (b) MPC based on PPC.

The control strategy of MPC techniques involves the following four steps:

1) The three-phase current and voltage of the BESS are measured, and the values of reference signals are obtained from the outer control loop.

2) The discrete-time model of the converter is used to predict the values of current or real/reactive powers in the next sampling interval $(k + 1)$ for each voltage vector according to Equations (14)–(16).

3) The cost function gC or gS based on Equations (17) and (18) is used to compute the errors between the reference and the predicted current or real/reactive powers for each voltage vector.

4) The minimum value of the cost function gives the minimum error between the reference and the measured signals. The voltage vector with respect to the minimum cost function is selected, and the corresponding switching state signals are generated to apply to the converter.

3. TEST MICROGRID

The test microgrid system (Figure 4) used in this study includes several components: A diesel generator, a consumer load, a wind generator, and two BESSs. Table 2 shows the parameters of the test microgrid system. In this study, the fixed-speed wind energy conversion system (WECS), a type of WECS [34], is used for simplicity. Two BESSs with different control strategies according to the operation mode of the microgrid, as shown in Table 3, are used. In the grid-connected mode, the voltage and frequency of the microgrid is set by the utility grid. Therefore, the main function of the BESS is to control the real and reactive powers. On the other hand, in the islanded mode, the microgrid is disconnected from the utility grid and controls its own frequency and voltage.

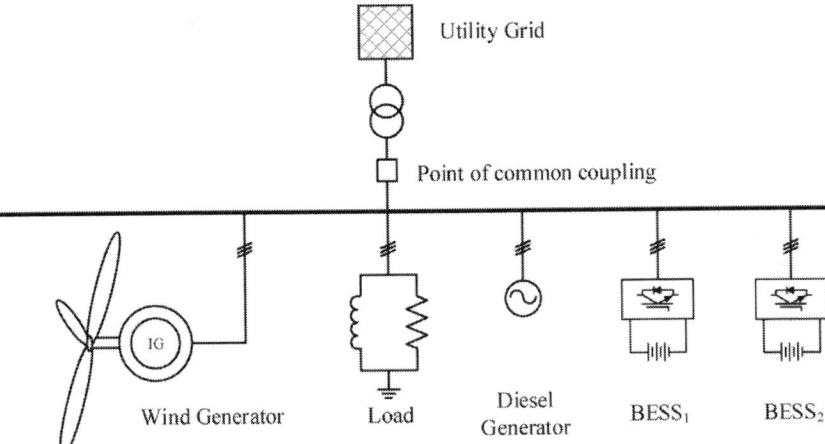

Figure 4. Configuration of microgrid.

Table 2. Parameters of test microgrid.

Components	Rating
Wind generator	150 kVA
BESS$_1$	450 kWh
BESS$_2$	200 kWh
Load	500 kW; 100 kVAR
Diesel generator	500 kVA
Mean wind speed	9 m/s
System frequency	60 Hz
Transformer	700 kVA; 6.6 kV/380 V

Table 3. Control strategies of BESSs.

Operation modes	BESS$_1$	BESS$_2$
Grid-connected	Tie-line powers at point of common coupling	Smoothing wind power
Islanded	Frequency control	Smoothing wind power
	Reactive power at point of common coupling	Voltage control

4. CONTROL PERFORMANCE OF MPC TECHNIQUES

4.1. Comparison of Control Performance of MPC and PI Control Techniques

The control performances of two MPC techniques according to the change in real power are compared to that of the PI control technique proposed in [35]. Tuning the PI parameters is an important factor for comparison. Several functions as well as linear analysis tools provided by MATLAB/Simulink are used for tuning. First, the function "getlinio" is used to obtain the linearized input/output of the plant. The linear approximation of the plant is estimated based on the linearized input/output by using the "linearize" function. Then, the linear analysis tool in Simulink is used to estimate the frequency response of a plant based on the linear approximation of the plant. Finally, the PID tuner in Simulink is used to automatically tune the PI parameters based on the frequency response estimation.

Figure 5 shows the simulation results of three types of control techniques. The real power changes from 0 to 50 kW at 1.0 s. The response of the PPC technique is clearly much quicker than that of other techniques. In the case of MPC based on PI in the outer and PCC in the inner control loops and PI control technique using PI regulators in the outer and inner control loops, the dynamic response is similar owing to the action of the PI controller in the outer control loop. Both MPC technique based on PI and PCC and PI technique show good reference tracking under the steady-state condition. However, the power ripple obtained by MPC technique is smaller than that obtained by PI control technique owing to PCC in the inner control loop in MPC technique. Figure 5 shows that MPC techniques can significantly improve the performance of a control system for BESSs in terms of the response time and power ripple.

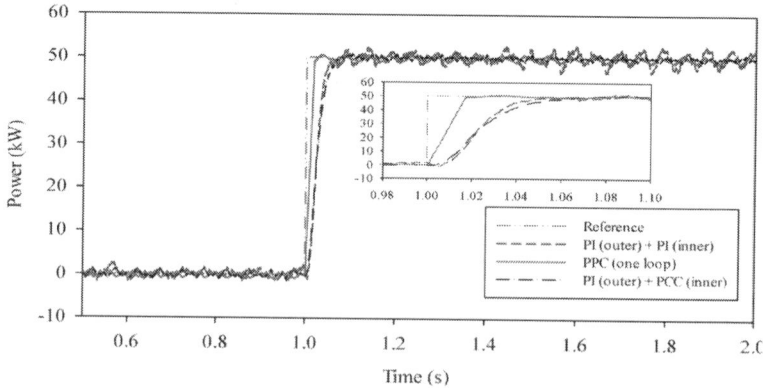

Figure 5. Response of different control techniques for change in reference power.

4.2. Effective Application of MPC Techniques to Microgrid Control

Table 4 shows the characteristics of the MPC and PI control techniques. Among these two MPC techniques, the PPC technique shows the best control performance; however, it can only be used for controlling the power. On the other hand, the MPC technique based on PI in outer and PCC in inner control loops is more flexible owing to the use of a PI regulator in the outer control loop; this technique can be used to control the power, frequency, and voltage. The ripple in case of both MPC techniques is smaller than that in case of the PI control technique.

Table 4. Characteristics of MPC and PI control techniques.

Characteristics	PI (outer) + PI (inner)	PI (outer) + PCC (inner)	PPC (one loop)
Ability to control	P/Q, f/v	P/Q, f/v	P/Q
Response time	Long	Long	Short
Ripple	Large	Small	Small

In this study, two BESSs with different functionalities are proposed to control the microgrid, as shown in Table 3. $BESS_1$ is used to control the power at the point of common coupling and the frequency in the islanded mode, in which case fast dynamic response under disturbances is required for the control system. Therefore, PPC-based MPC is suitable for application to $BESS_1$ because its control performance shows the shortest response time compared to other cases. Furthermore, $BESS_2$ is used for handling fluctuations in wind power in both grid-connected and islanded modes. Thus, the control performance of the MPC technique based on PI control in the outer control loop and PCC control in the inner control loop is suitable for $BESS_2$ owing to gradual fluctuations in wind power. The microgrid voltage is controlled by $BESS_2$ and the frequency, by $BEES_1$ and $BESS_2$ through the improved frequency droop control scheme.

5. SIMULATION RESULTS

5.1. Control Microgrid in Grid-Connected Mode

BESSs can operate in the charging or discharging mode. Therefore, they can reduce the fluctuations in wind power through effective compensation. Figure 6 shows the action of $BESS_2$ in terms of smoothing the wind power. In the case of $BESS_2$, the MPC technique based on PI control in the outer control loop and PCC in the inner current control loop is applied as the control system. This figure shows that the wind power fluctuations can be reduced significantly by effectively charging or discharging $BESS_2$. Both the MPC and the PI control techniques show good results from the viewpoint of smoothing the wind power. However, the power ripple in case of the MPC technique is much smaller than that in case of the PI control technique.

On the other hand, BESS₁ based on the PPC technique controls the power at the point of common coupling. In this study, it is assumed that the real power at the point of common coupling is maintained at zero. Figure 7shows the simulation result. At 10 s, an additional load of 100 kW is connected to the microgrid. Therefore, BESS₁increases its real power to maintain the power at zero. The subfigure of Figure 7 shows that the response of the MPC technique is slightly quicker than that of the PI control technique. Additionally, the ripples of the BESS power when using the MPC technique is smaller than that of PI control technique. Both the MPC and the PI control techniques show good performance for controlling the power at the point of common coupling.

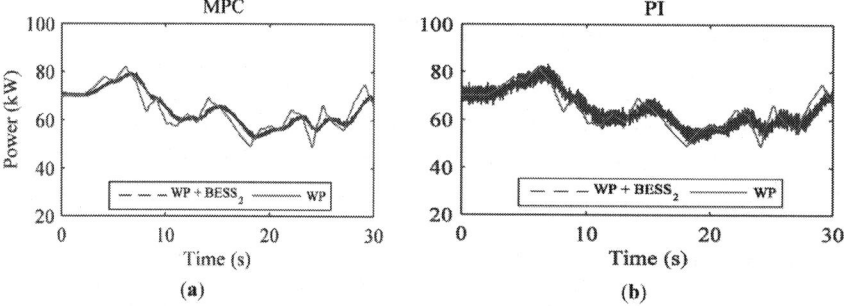

Figure 6. Smoothened wind power: (**a**) MPC technique; (**b**) PI technique.

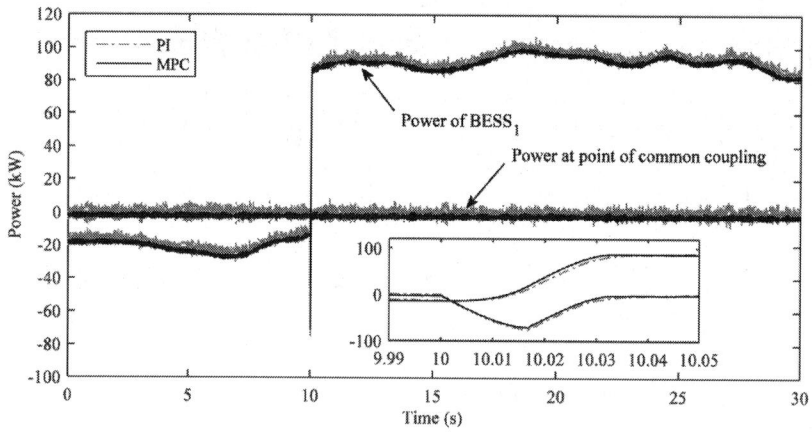

Figure 7. Real power at point of common coupling and real power of BESS₁.

5.2. Control Microgrid in Islanded Mode

In the islanded mode, the microgrid frequency is controlled by BESS₁, and the microgrid voltage is controlled by BESS₂. Figure 8 and Figure

9 respectively show the frequency and voltage of the microgrid. Both the MPC and the PI control techniques can stably control the frequency and voltage of the microgrid. However, as shown inFigure 8, the frequency response under the MPC technique is quicker than that under the PI control technique. Moreover, Figure 9 shows the microgrid voltage. Obviously, the performance of the MPC techniques is much better than that of the PI control technique. The voltage ripple in the case of the MPC technique is much smaller than that in the case of the PI control technique.

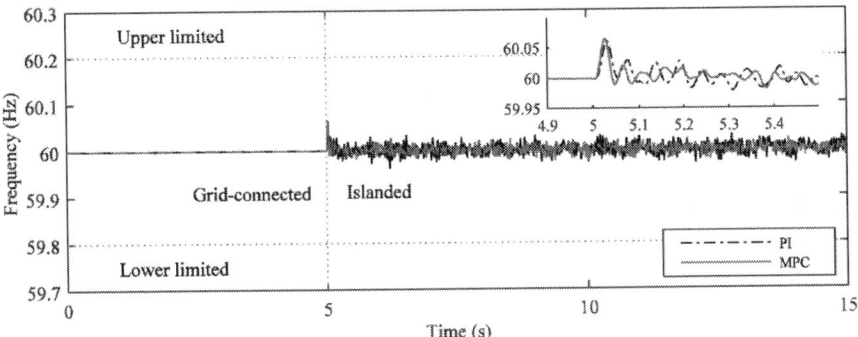

Figure 8. Frequency of microgrid system.

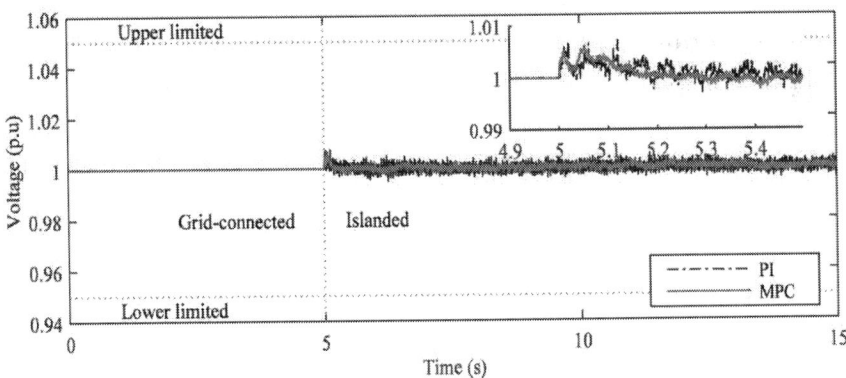

Figure 9. Voltage of microgrid system.

Additionally, the output voltage spectra generated by the converter is one of the important factors. Figure 10shows a comparison of the voltage spectra of the MPC and PI control techniques. As shown in Figure 10b, the frequency spectrum generated using the PI control technique is concentrated around the carrier frequency owing to PWM. For comparison, Figure 10a shows the frequency spectrum obtained by MPC. The reduction of the switching frequency of the converter is implemented in the cost function of MPC as a secondary control objective to reduce the power losses of converters. Figure 10 shows that the average switching frequency (fs) obtained by MPC is slightly lower than that obtained by the PI control technique. Moreover, the MPC technique shows significantly lower THD than the PI control technique.

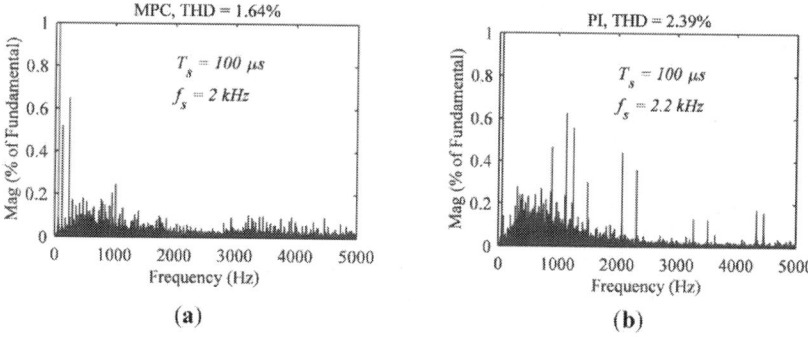

Figure 10. Load voltage spectrum and THD: (**a**) MPC technique; (**b**) PI technique.

6. CONCLUSIONS

This study discusses the effective application of two types of MPC techniques to BESSs for microgrid control: MPC based on PPC and MPC based on PI control in the outer control loop and PCC in the inner current control loop. In addition, PI control using a PI regulator in the outer and inner control loops for BESS was compared to these two types of MPC techniques. A reduction switching frequency is implemented in the cost function to reduce the power losses of converters. The simulation results show that the response time, power ripples, and frequency spectrum could be improved significantly by using MPC techniques. Both the average switching frequency and the THD obtained by using MPC techniques were lower than those obtained by using PI control. Using MPC based on PI control in the outer and PCC in the inner control loops did not improve the response time under power changing compared to PI control; however, it could significantly improve the power and voltage ripples under the steady-state condition. Moreover, using PPC-based MPC could reduce the response time under power changing compared to other control techniques. Therefore, in microgrids with multiple BESSs, the PPC-based MPC technique should be applied for BESSs that control the power at the point of common coupling and the frequency of the microgrid, and an MPC technique based on PI in the outer control loop and PCC in the inner control loop should be applied for BESSs that play the role of smoothing wind power fluctuations. Besides, in case of microgrids with a BESS, PCC-based MPC technique should be a suitable alternative for the BESS owing to its flexible characteristic. MPC technique is easy to implement and it can eliminate the tuning controller parameters effort that has to be done in the PI technique. Furthermore, various control objectives can be included in the MPC strategies.

In the future, we plan to include additional control variables such as considering the state of charge of the battery and coordination control of multiple ESSs in the MPC algorithm.

ACKNOWLEDGMENTS

This work was supported by the Power Generation & Electricity Delivery Core Technology Program of the Korea Institute of Energy Technology Evaluation and Planning (KETEP), granted financial resource from the Ministry of Trade, Industry & Energy, Republic of Korea. (No. 20141020402350).

AUTHOR CONTRIBUTIONS

The paper was a collaborative effort between the authors. The authors contributed collectively to the theoretical analysis, modeling, simulation, and manuscript preparation.

REFERENCES

1. Mahmoud, M.S.; Hussain, S.A.; Abido, M.A. Modeling and control of microgrid: An overview. *J. Frankl. Inst.* **2014**, *351*, 2822–2859.
2. Hatziargyriou, N.D. Microgrids. *IEEE Power Energy* **2008**, *6*, 26–29.
3. Olivares, D.E.; Mehrizi-Sani, A.; Etemadi, A.H.; Canizares, C.A.; Iravani, R.; Kazerani, M.; Hajimiragha, A.H.; Gomis-Bellmunt, O.; Saeedifard, M.; Palma-Behnke, R.; *et al.* Trends in microgrid control. *IEEE Trans. Smart Grid* **2014**, *5*, 1905–1919.
4. Kim, H.-M.; Lim, Y.; Kinoshita, T. An intelligent multiagent system for autonomous microgrid operation. *Energies* **2012**, *5*, 3347–3362.
5. Kim, H.-M.; Kinoshita, T. A multiagent system for microgrid operation in the grid-interconnected mode. *J. Electr. Eng. Technol.* **2010**, *2*, 246–254.
6. Molina, M.G.; Mercado, P.E. Power flow stabilization and control of microgrid with wind generation by superconducting magnetic energy storage. *IEEE Trans. Power Electron.* **2011**, *26*, 910–922.

7. Inthamoussou, F.A.; Pegueroles-Queralt, J.; Bianchi, F.D. Control of a supercapacitor energy storage system for microgrid applications. *IEEE Trans. Energy Conver.* **2013**, *28*, 690–697.

8. Islam, F.; Al-Durra, A.; Muyeen, S.M. Smoothing of wind farm output by prediction and supervisory-control-unit-based FESS. *IEEE Trans. Sustain. Energy* **2013**, *4*, 925–933.

9. Nguyen, T.-T.; Yoo, H.-J.; Kim, H.-M. A flywheel energy storage system based on a doubly fed induction machine and battery for microgrid control. *Energies* **2015**, *8*, 5074–5089.

10. Chang, X.; Li, Y.; Zhang, W.; Wang, N.; Xue, W. Active disturbance rejection control for a flywheel energy storage system. *IEEE Trans. Ind. Electron.* **2015**, *62*, 991–1001.

11. Li, X. Fuzzy adaptive kalman filter for wind power output smoothing with battery energy storage system.*IET Renew. Power Gener.* **2012**, *6*, 340–347.

12. Li, X.; Hui, D.; Lai, X. Battery energy storage station (BESS)-based smoothing control of photovoltaic (PV) and wind power generation fluctuations. *IEEE Trans. Sustain. Energy* **2013**, *4*, 464–476.

13. Lawder, M.T.; Suthar, B.; Northrop, P.W.C.; De, S.; Hoff, C.M.; Leitermann, O.; Crow, M.L.; Santhanagopalan, S.; Subramanian, V.R. Battery energy storage system (BESS) and battery management system (BMS) for grid-scale applications. *Proc. IEEE* **2014**, *102*, 1014–1030.

14. Duran, M.J.; Prieto, J.; Barrero, F.; Toral, S. Predictive current control of dual three-phase drives using restraned search techiniques. *IEEE Trans. Ind. Electron.* **2011**, *58*, 3253–3263.

15. Kouro, S.; Cortés, P.; Vargas, R.; Ammann, U.; Rodríguez, J. Model predictive control—A Simple and powerful method to control power converter. *IEEE Trans. Ind. Electron.* **2009**, *56*, 1826–1838.

16. Miranda, H.; Cortés, P.; Yuz, J.I.; Rodríguez, J. Predictive Torque control of induction machines based on state-space models. *IEEE Trans. Ind. Electron.* **2009**, *56*, 1916–1924.

17. Morel, F.; Xuefang, L.S.; Retif, J.M.; Allard, B.; Buttay, C. A comparative study of predictive current control schemes for a permanent-magnet synchronous machine drive. *IEEE Trans. Ind. Electron.* **2009**, *56*, 2715–2728.

18. Bolognani, S.; Peretti, L.; Zigliotto, M. Design and implementation of model predictive control for electrical motor drives. *IEEE Trans. Ind. Electron.* **2009**, *56*, 1925–1936.

19. Cortés, P.; Rodríguez, J.; Antoniewicz, P.; Kazmierkowski, M. Direct power control of an AFE using predictive control. *IEEE Trans. Power Electron.* **2008**, *23*, 2516–2523.

20. Vargas, R.; Rodríguez, J.; Ammann, U.; Wheeler, P.W. Predictive current control of an induction machine fed by a matrix converter with reactive power control. *IEEE Trans. Ind. Electron.* **2008**, *55*, 4362–4371.

21. Abad, G.; Rodriguez, M.A.; Poza, J. Three-level NPC converter based predictive direct power control of the doubly fed induction machine at low constant switching frequency. *IEEE Trans. Ind. Electron.* **2008**, *55*, 4417–4429. Torreglosa, J.P.; Garcia, P.; Femadez, L.M.; Jurado, F. Predictive control for the energy management of a fuel-cell-battery-supercapacitor tramway. *IEEE Trans. Ind. Electron.* **2014**, *10*, 276–285.

22. Hredzak, B.; Agelidis, V.G.; Jang, M. A model predictive control system for a hybrid battery-ultracapacitor power source. *IEEE Trans. Power Electron.* **2014**, *29*, 1469–1479.

23. Akter, M.P.; Mekhilef, S.; Tan, N.M.L.; Akagi, H. Model predictive control of bidirectional AC-DC converter for energy storage system. *J. Electr. Eng. Technol.* **2015**, *10*, 165–175.

24. John, T.; Wang, Y.; Tan, K.T.; So, P.L. Model predictive control of distributed generation inverter in a microgrid. In Proceedings of the 2014 IEEE Innovative Smart Grid Technologies (ISGT Asia), Kuala Lumpur, Malaysia, 20–23 May 2014; pp. 657–662.

25. Jafari, H.; Mahmodi, M.; Rastegar, H. Frequency control of micro-grid in autonomous mode using model predictive control. In Proceedings of the 2012 IEEE Iranian Conference on Smart Grids (ICSG), Tehran, Iran, 24–25 May 2012; pp. 1–5.

26. Naeiji, N.; Hamzeh, M.; Rahimi Kian, A. A modified model predictive control method for voltage control of an inverter in islanded microgrids. In Proceedings of the 2015 IEEE Power Electronics, Drives Systems & Technologies Conference (PEDSTC), Tehran, Iran, 3–4 February 2015; pp. 555–560.

27. Han, J.; Solanki, S.K.; Solanki, J. Coordinated predictive control of a wind-battery microgrid system. *IEEE J. Emerg. Sel. Top. Power Electron.* **2013**, *1*, 296–305.

28. Rodríguez, J.; Cortés, P. *Predictive Control of Power Converter and Electrical Drives*; John Wiley & Sons: West Sussex, UK, 2012.

29. Xia, C.; Wang, M.; Song, Z.; Liu, T. Robust model predictive current control of three-phase voltage source PWM rectifier with online disturbance observation. *IEEE Trans. Ind. Informat.* **2012**, *8*, 459–471.

30. Antoniewicz, P.; Kazmierkowski, M.P. Virtual-flux-based predictive direct power control of AC/DC converters with online inductance estimation. *IEEE Trans. Ind. Electron.* **2008**, *55*, 4381–4390.

31. Rivera, M.; Yaramasu, V.; Rodriguez, J.; Wu, B. Model predictive current control of two-level four-leg inverters—Part II: Experimental implementation and validation. *IEEE Trans. Power Electron.* **2013**, *28*, 3469–3478.

32. Natesan, C.; Ajithan, S.; Mani, S.; Kandhasamy, P. Applicability of droop regulation technique in microgrid—A survey. *Eng. J.* **2014**, *18*, 23–35.

33. Wu, B.; Lang, Y.; Zargari, N.; Kouro, S. *Power Conversion and Control of Wind Energy Systems*, 1st ed.; Wiley-IEEE Press: Hoboken, NJ, USA, 2011; pp. 153–170.

34. Yoo, H.-J.; Kim, H.-M.; Song, C.-H. A coordinated frequency control of lead-acid BESS and Li-ion BESS during islanded microgrid operation. In Proceedings of the 2012 IEEE Vehicle Power and Propulsion Conference (VPPC), Seoul, Korea, 9–12 October 2012; pp. 1453–1456.

CHAPTER 3
Model Predictive Control Strategies for Batch Sugar Crystallization Process

Luis Alberto Paz Suárez[1], Petia Georgieva[1] and Sebastião Feyo de Azevedo[1]

[1] Faculty of Engeneering, University of Porto, Portugal

[2] Institute of Electronic Engineering and Telematics of Aveiro, Portugal

1. INTRODUCTION

The industrial processes are governed generally by general principles of the physics and chemistry. With the aid of data acquisition systems supported in microprocessor it is possible to obtain real data of the industrial process, that it characterizes in detail his dynamics and input-output dependency. Several methods of identification allow, from these data, to obtain linear and nonlinear models of these processes (Rossiter, 2003; Morari, 1994); which are the base to predict the process behaviour within all the family of the model based predictive controllers (MPC).

Diverse algorithms MPC have demonstrated its effectiveness in those control loops characterized by strong nonlinearities, difficult dynamic, inverse answers and great delay; that they are generally those of greater influence in the final product quality and the process efficiency (Allgöwer et al., 2004; Qin & Badgwell 2003).

One of the most important steps in the implementation of a MPC is just the obtaining of the model that can predict with reliability the future behaviour of the controlled variable, like answer to a predefined optimized control action (Rawlings 2000). This work applies two kind of MPC: (i) Classical Model-Based Predictive Control and (ii) Neural Network Model Predictive Control (NNMPC).

The classical MPC strategy uses a discrete model obtained from general phenomenological model of the feed-batch crystallization process, consisting of mass, energy and population balance. The NNMPC strategy uses to obtain a neural network, the training algorithms proposed in the Neural Network Toolbox of MatLab (version 7.04) (Bemporad et al., 2005).

In this particular case it is analyzed a fed-batch sugar crystallization process, in this process there is abundant information, detailed mathematical models and real industrial data. (Chorão, 1995; Feyo de Azevedo & Gonçalves

1988; Georgieva et al., 2003). This fact motivated the use of the neural networks to model the process and to propose a neural network MPC (NNMPC) that considers the process like a gray box, of which has input-output information and the historical experience of he process behaviour.

2. BATCH SUGAR CRYSTALLIZATION PROCESS

2.1. General Description

The operation of crystallization is applied in the sugar industry to obtain the sucrose dissolved in the extracted juice of the sugar cane or the sugar beet basically.

Typical industrial fed-batch evaporative sugar crystallization is performed in a vacuum pan crystallizer. The reactor has a cylindrical form with volume that can vary between 20-60 m3. The feed system is usually equipped with an extra water input to dilute the sugar solution if necessary. The heat transfer system is a calandria type, to permit the heat interchange between steam and suspension. The vacuum pressure in the pan is generated by the contact barometric condenser and the pan is equipped with a mechanical agitator to keep the suspension homogeneous. The operation is conducted in a fed-batch mode with an average duration of a cycle about 90 minutes.

Sugar crystallization occurs through the mechanisms of nucleation, growth and agglomeration. In the course of production, the crystallization phenomenon is driven by two mechanisms (Jancic & Grootscholten, 1984): i) mass transfer from dissolved sucrose to crystal surface and ii) heat transfer in the calandria. Shortly before the grain setting and continuing during the beginning of the crystallization phase, the available crystalline surface to deposit the molecule of sucrose is much smaller than the mass of dissolved sucrose. During this period the evaporation rate is high, the crystal area/mass of crystallized sucrose rate is very low, therefore the process is driven by the mass transfer. The supersaturation tends to increase and if not controlled, it often achieves the undesirable zone of secondary crystal nucleation. Later on, when the total crystal area and the crystallization capacity increases, the crystal area/mass of crystallized sucrose rate gets high and the process is driven by the heat transfer.

The process objectives are to maximize the speed of crystal growth, keeping high the produced sugar quality and minimizing the costs and losses. These objectives must be fulfilled without occurrence of secondary nucleation or agglomeration. The sugar quality is evaluated by the particle size distribution (PSD) at the end of the process which is quantified by two parameters - the final average (in mass) particle size (MA) and the final coefficient of particle variation (CV). The main challenge of the sugar production is the large batch to batch variation of the final PSD. This lack of process repeatability is caused mainly by improper control policy and results in product recycling and loss increase. The sugar production is heuristically operated, and while the traditionally applied PI(D) controllers are still the preferred solutions they usually lead to energy and material loss that can easily be reduced if an optimized operation policy is implemented. These problems constitute the main motivation for the operation strategy formulated in the next section.

2.2. Crystallization Model

The general phenomenological model of the fed-batch crystallization process consists of mass, energy and population balances, including the relevant kinetic rates for nucleation, linear growth and agglomeration (Simoglou et al., 2005). While the mass and energy balances are common expressions in many chemical process models, the population balance is related with the crystallization phenomenon, which is still an open modelling problem. The Appendix A shows a detailed phenomenological model for crystallization process.

2.3. Problem Formulation

The final values of the crystal size distribution function (CSD) parameters: mass averaged crystal size (MA) and coefficient of variation (CV) are the best indicators of the quality and efficiency of the crystallization process. The direct measurement and control of these parameters are very difficult to make actually, in fact there are no references of its industrial implementation. The most used solution in the sugar industry consists of establishing a strategy that manipulate other variables; which allows to arrive at the end of the process with acceptable values in the CSD parameters.

The batch operation imposes to the process frequent operational changes that depend of: the quality of the raw material, disturbances in the work conditions and market demand changes. The previous problem, the nonlinearities and the restrictions imposed to the process motivated the use a nonlinear MPC (NMPC).

When a NMPC algorithm is applied, the first challenge consists of obtaining the model to use, which must be viable and trustworthy. Although the sugar crystallization process has been studied in depth and efficient mathematical models exist to represent it, these must be validated and be fit before its application in a NMPC algorithm, which will cause frequent updates if the process is batch.

Like an alternative, in this work it is tried to demonstrate the efficiency that has the use of the neuronal networks in a NMPC, where the neural networks could be trained from industrial data with the input-output answer of the process.

3. PROBLEM SOLUTION

Sugar production is characterized by strongly non-linear and non-stationary dynamics and goes naturally through a sequence of relatively independent stages: charging, concentration, seeding, setting the grain, crystallization (the main phase), tightening and discharge (Georgieva et al., 2003). Therefore the operation strategy is formulated as a cascade of individual control loops for each of the stages (Fig. 1). The feedback control policy is based on measurements of the flowrate, the temperature, the pressure, the stirrer power and the supersaturation (by a refractometer). Measurements of these variables are usually available for a conventional crystallizer.

3.1. Operation Strategy

Sugar production is still a very heuristically operated process, with classical proportional integral and eventually derivative (PID) controllers being the most typical solution. The different phases of the sugar production are comparatively independent and moved by distinct driving forces, thus a single controller can hardly be effective for the complete process. Instead, individual controllers for each stage where it seems appropriate, was the adopted framework (Fig. 1). See Table 1 for more details on the formulated operation strategy.

In the present study, the control actions are performed by manipulating the valves of the liquor/syrup feed flowrates (Ff) and the steam flowrate (Fs), while the volume of massecuite (Vm), the supersaturation (S) and the current of the agitator (IA) are the controlled variables. This choice is completely inspired by the industrial practice in several refineries.

Charging (stage 1): During the first stage the crystallizer is fed with liquor until it covers approximately 40 % of the vessel height. The process starts with vacuum pressure of around 1 bar (equal to the atmospheric pressure) and reduces it up to 0.23 bar. When the vacuum pressure reaches 0.5 bar, the feed valve is completely open such that the feed flowrate is kept at its maximum value. When the liquor covers 40 % of the vessel height, the feed valve is closed and the vacuum pressure needs some time to stabilize around the value of 0.23 bar before the concentration stage starts.

Concentration (stage 2): Once the vacuum pressure stabilizes, the stirrer is switched on and the concentration begins. In order to guarantee unperturbed operation of the barometric condenser and the steam production boiler, the steam flowrate must increase slowly (from 0 to 2 kg/s, in two minutes approximately). The concentration of the dissolved sucrose by evaporation under vacuum results in volume reduction. However, for technological reasons, the minimum suspension level of the pan must be above the calandria. Therefore a feed flowrate action is required to control the level (the volume) of the pan and this constitutes the *first control loop*. In this stage, the supersaturation increases rapidly (at about a rate of 0.025 per *min.*). When it reaches a value of 1.06, the feeding is stopped and the steam flowrate is reduced slowly to 1.4 kg/s, with the same speed as it was increased. The concentration stage is over when the supersaturation reaches the value of 1.11.

Seeding (stage 3): At this moment seed crystals are introduced into the pan to provoke crystallization. This stage is rather unstable and to prevent seed crystals from dissolution in the liquor, the feed valve must be closed and the steam flowrate kept at its minimum for a short period (about 2 *min.*). Keeping these conditions unchanged contributes to the formation of the grain and is also important for the final crystal size distribution. The supersaturation continues naturally to increase but usually no control action is required.

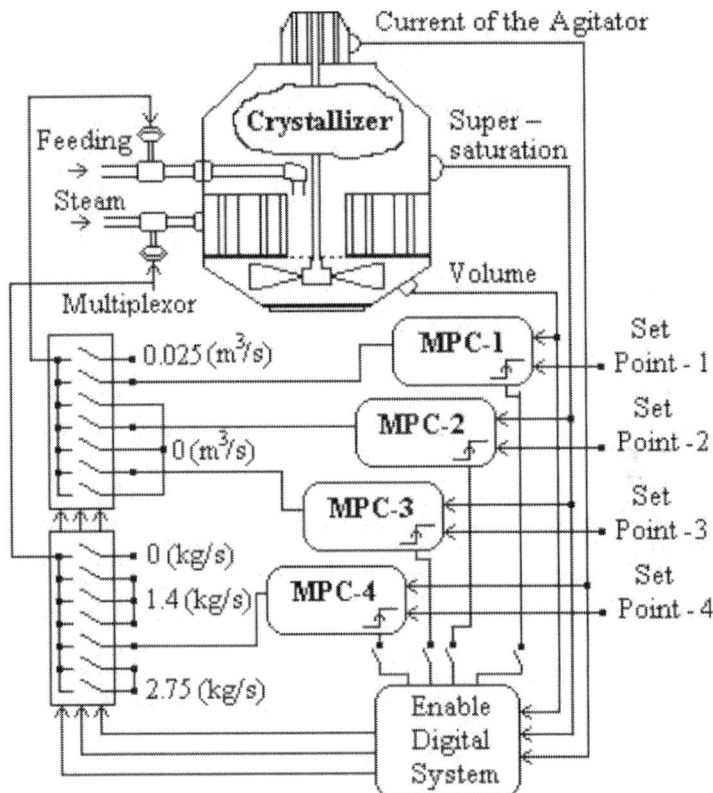

Figure 1. cascade mpc control - strategy

Crystallization with liquor (stage 4): During this stage the supersaturation is first controlled by a proper feeding to be around a set point of 1.15. This constitutes the *second control loop*. At the beginning of this stage, the mass transfer is the driving crystallization force, the crystallization rate increases and the controller usually reduces the feed flowrate to maintain the reference value of the supersaturation. When all liquor quantity is introduced, the feeding is stopped and the supersaturation is now kept at the same set point of 1.15 by the steam flowrate as the manipulated variable. This constitutes the *third control loop*. The heat transfer is now the driving crystallization force. A typical problem of this control loop is that at the end of this stage the steam flowrate achieves its maximum value of 2.75 *kg/s* but it is not sufficient to keep the supersaturation at the same reference value therefore a reduction of the set point is required. The stage is over when the stirrer power reaches the value 20.5 *A*.

Table 1. Summary of the sugar crystallization operation strategy.

Stage	Action	Control
Charge	The steam valve is closed and the stirrer is off. The vacuum pressure changes from 1 to 0.23 *bar*. The vacuum pressure reaches 0.5 *bar*, feeding starts with max rate. Liquor covers 40 % of the vessel height.	No control The feed valve is completely open
Concentration	The vacuum pressure stabilizes around 0.23 *bar*. The stirrer is on. The volume is kept constant. The steam flowrate increases to 2 *kg/s* The supersaturation reaches 1.06, the feeding is closed, the steam flowrate is reduced to 1.4 *kg/s*	*Control loop 1* Controlled variable: Volume; Manipulated variable: liquor feed flowrate
Seeding and setting the grain	The supersaturation reaches 1.11. Seed crystals are introduced. The steam flowrate is kept at the minimum for two minutes.	No control The feed valve is closed
Crystallization with liquor (phase 1)	The steam flowrate is kept around 1.4 *kg/s*. The supersaturation is controlled at the set point 1.15.	*Control loop 2* Controlled variable: supersaturation Manipulated variable: liquor feed flowrate
Crystallization with liquor (phase 2)	The volume of crystallizer reaches ≈ 22 m^3. The feed valve is closed. The supersaturation is controlled at the set point 1.15. The stirrer power reaches 20.5 *A*.	*Control loop 3* Controlled variable: supersaturation Manipulated variable: steam flowrate
Crystallization with syrup	The steam flowrate is kept around the maximum of 2.75 *kg/s* (*hard constraint*). The volume fraction of crystals is kept at the set point 0.45. The volume reaches its maximum value (30 m^3) The feed valve is close.	*Control loop 4* Controlled variable: volume fraction of crystals. Manipulated variable: syrup feed flowrate
Tightening	The stirrer power reaches the maximum value of 50 A (*hard constraint*). The steam valve is closed. The stirrer and the barometric condenser are stopped.	No control

Crystallization with syrup (stage 5): A stirrer power of 20.5*A* corresponds to a volume fraction of crystals equal to 0.4. At this moment the feed valve is reopened, but now a juice with less purity (termed syrup) is introduced into the pan until the maximum volume (30 m^3) is reached. The control objective is to maintain the volume fraction of crystals around the set point of 0.45 by a proper syrup feeding. This constitutes the *fourth control loop*.

Tightening (stage 6): Once the pan is full the feeding is closed. The tightening stage consists principally in waiting until the suspension reaches the reference consistency, which corresponds to a volume fraction of crystals equal to 0.5. The supersaturation is not a controlled variable at this stage because due to the current conditions in the crystallizer, the crystallization rate is high and it prevents the supersaturation of going out of the metastable zone. The stage is over when the stirrer power reaches the maximum value of 50 A. The steam valve is closed, the water pump of the barometric condenser and the stirrer are turned off. Now the suspension is ready to be unloaded and centrifuged.

4. MODEL BASED PREDICTIVE CONTROL

The term model-based predictive control (MPC) does not refer to a particular control method, instead it corresponds to a general control approach (Rossiter, 2003). The MPC concept, introduced in late seventies, nowadays has evolved to a mature level and became an attractive control strategy implemented in a variety of process industries (Camacho & Bordons, 2004). The main difference between the MPC configurations is the model used to predict the future behavior of the process or the implemented optimization procedure. First the MPC based on linear models gained popularity (Morari, 1994) as an industrial alternative to the classical proportional-integral-derivative (PID) control and later on nonlinear cases as reactive distillation columns (Balasubramhanya & Doyle, 2000) and polymerization reactors (Seki et al., 2001) were reported as successfully MPC controlled processes.

4.1. Classical model based predictive control

The main difference between MPC configurations is the model used to predict the future behaviour of the process and the optimization procedure. Nonlinear model predictive control (NMPC) is an optimisation-based multivariable constrained control technique that uses a nonlinear dynamic model for the prediction of the process outputs (Qin & Badgwell, 2003). At each sampling time k the model predicts future process responses to potential control signals over the prediction horizon (H_p). The predictions are supplied to an optimization procedure, to determine the values of the control action over a specified control horizon (H_c) that minimizes the following performance index:

$$\min_{u_{min} \leq [u_c(k), u_c(k+1), \dots u_c(H_c)] \leq u_{max}} J = \lambda_1 \sum_{k=1}^{H_p} (y_r(k) - \hat{y}(k))^2 - \lambda_2 \sum_{k=1}^{H_c} (u_c(k-1) - u_c(k-2))^2$$

(1)

Subject to the following constrains

$$u_{min} \leq u_c \leq u_{max}$$

(2)

$$\Delta u_{min} \leq \Delta u \leq \Delta u_{max}$$

(3)

$$y_{\min} \leq y_p \leq y_{\max}$$

(4)

Where $u_{\min}$ and $u_{\max}$ are the limits of the control inputs, $\Delta u_{\min}$ and $\Delta u_{\max}$ are the minimum and the maximum values of the rate-of-change of the inputs and min $y_{\min}$ and $y_{\max}$ are the minimum and maximum values of the process outputs.

H_p is the number of time steps over which the prediction errors are minimized and the control horizon H_c is the number of time steps over which the control increments are minimized, y_r is the desired response (the reference) and $\hat{y}$ is the predicted process output (Diehl et al., 2002). $u_c(k), u_c(k+1), u_c(H_c)$ are tentative future values of the control input, which are parameterized as peace wise constant. The length of the prediction horizon is crucial for achieving tracking and stability. For small values of H_p the tracking deteriorates but for high H_p values the bang-bang behavior of the process input may be a real problem. The MPC controller requires a significant amount of on-line computation, since the optimization (1) is performed at each sample time to compute the optimal control input. At each step only the first control action is implemented to the process, the prediction horizon is shifted or shrunk by usually one sampling time into the future, and the previous steps are repeated (Rossiter, 2003). λ_1 and λ_2 are the output and the input weights respectively, which determine the contribution of each of the components of the performance index (1).

4.2. Neural network model predictive control

The need for neural networks arises when dealing with non-linear systems for which the linear controllers and models do not satisfy. Two main achievements contributed to the increasing popularity of the NNs: (i) The proof of their universal approximation properties and the development of suitable algorithms for NN training as the backpropagation and (ii) The adaptation of the Levenberg-Marquard algorithm for NN optimization.

The most used NN structures are Feedforward networks (FFNN) and Recurrent (RNN) ones. The RNNs offer a better suited tool for nonlinear system modelling and is implemented in this work (Fig.2). The Levenberg-Marquard (LM) algorithm was preferred as the training method due to its advantages in terms of execution time and robustness. Since the LM algorithm requires a lot of memory, a powerful (in terms of memory) computer is the main condition for successful training. In order to solve the problem of several local minima, that is typical for all derivative based optimization algorithms (including the LM method), we have repeated several time the optimization specifying different starting points.

The individual stages of the crystallization process are approximated by different RNNs of the type shown in Fig. 2. Tangent sigmoid hyperbolic activation functions are the hidden computational nodes (Layer 1) and a linear function is located at the output (Layer 2). Each NN has two vector inputs (r and p) formed by past values of the process input and the NN output respectively.

The architecture of the NN models trained to represent different process stages is summarized as follows:

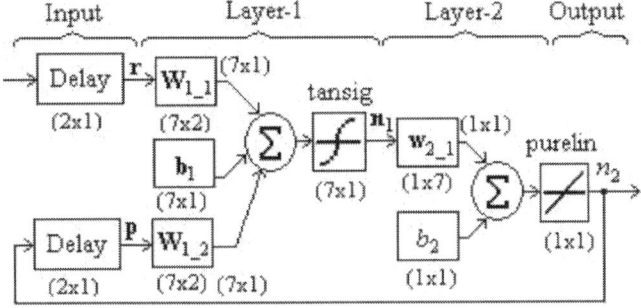

Figure 2. Neural network architecture

$$u_{NN} = [r, p] = [u_c(k-1), u_c(k-2), y_{NN}(k-1), y_{NN}(k-2)] \tag{5}$$

$$x = W_{11}r + W_{12}p + b_1 \tag{6}$$

$$n_1 = \left(e^x - e^{-x}\right) / \left(e^x + e^{-x}\right) \tag{7}$$

$$n_2 = w_{21}n_1 + b_2 \tag{8}$$

Where $W_{11} \in R^{m \times 2}$, $W_{12} \in R^{m \times 2}$, $w_{21} \in R^{1 \times m}$, $b_1 \in R^{m \times 1}$, $b^2 \in R$ are the network weights (in matrix form) to be adjusted during the NN training, m is the number of nodes in the hidden layer.

Since the objective is to study the influence of the NNs on the controller performance, a number of NN models is considered based on different training data sheets.

- **Case 1 (Generated data):** Randomly generated bounded inputs (u_i) are introduced to a simulator of a general evaporative sugar crystallization process introduced in Georgieva et al., 2003. It is a system of nonlinear differential equations for the mass and energy balances with the operation parameters computed based on empirical relations (for no stationary parameters) or keeping constant values (for stationary parameters). The simulator responses are recorded (y_i) and the respective mean values are computed ($u_{i,mean}, y_{i, mean}$). Then the NN is trained supplying as inputs $u_i - u_{i,mean}$ and as target outputs $y_i - y_{i,mean}$.

- **Case 2: Industrial data:** The NN is trained with real industrial data. In order to extract the underlying nonlinear process dynamics a prepossessing of the initial industrial data was performed. From the complete time series corresponding to the input signal of one stage only the portion that really excites the process output of the same stage is extracted. Hence, long periods of constant (steady-state) behavior are discarded. Since, the steady-state periods for normal operation are usually preceded by transient intervals, the data base constructed consists (in average) of 60-70% of transient period data. A number of sub cases are considered.

  - o **Case 2.1:** Industrial data of two batches is used for NN training.
  - o **Case 2.2:** Industrial data of four batches is used for NN training.
  - o **Case 2.3:** Industrial data of six batches is used for NN training.

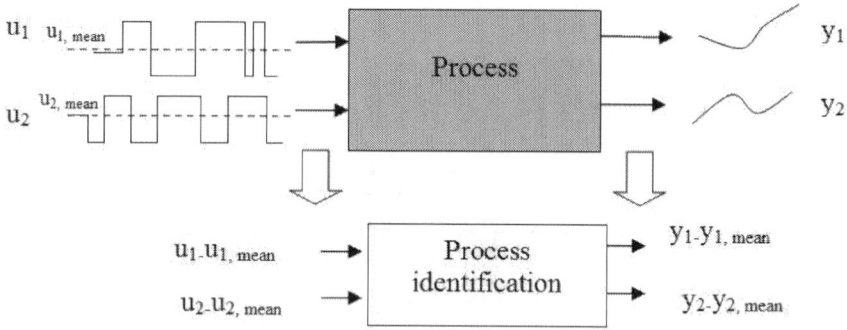

Figure 3. Case1: NN data generation

4.3. Selection of Mpc Parameters: H_p, H_c, Λ_2

The choice of H_p is related with the sampling period (Δt) of the digital control implementation, which in its turn is a function of the settling time t_s (the time before entering into the 5% around the set-point) of the closed loop system. As a rule of thumb, it is suggested Δt to be chosen at least 10 times smaller than t_s, (Soeterboek, 1992). Hence, the prediction horizon can be chosen as $H_p =$ round-to-integer($t_s/\Delta t$). It is well known that the smaller the sampling time, the better can a reference trajectory be tracked or a disturbance rejected. However, choosing a small sampling time yields a large prediction horizon. In order to compute the optimal control input, the optimization (1) is performed at each sampling time, therefore MPC controller requires a significant amount of on-line computation. This can cause problems related with large amount of computer memory required and additional numerical problems due to the large prediction horizon. The introduction of the ET MPC as in (7) serves as a compromise between these conflicting issues and reduces significantly the computational efforts.

Parameters λ_1 and λ_2 determine the contribution (the weight) of each term of the performance index, the output error (e) and the control increments (Δu). In this work the parameter λ_1 is set to the normalized value of 1, while the choice of λ_2 is based on the following empirical expression:

$$\left(u_{max} - u_{min}\right)^2 \cdot \lambda_2 = e_{max} \cdot P/100 \tag{9}$$

where P defines the desired contribution of the second term in (1) ($0\% \leq P \leq 100\%$) and

$$e_{max} = max\left(\left(ref - y_{max}\right)^2, \left(ref - y_{min}\right)^2\right) \tag{10}$$

The intuition behind (9-10) is to make the two terms of (1) compatible when they are not normalized and to overcome the problem of different numerical ranges for the two terms. Table 2 summarize the set of MPC parameters used in the four control loops define in the section 3.

Table 2. MPC design parameters for the control loops define in Table 1

Control loop (CL)	t_s (s) settling time	Δt (s) sampling period	H_p prediction horizon	H_c control horizon	λ_2 weight	Controlled variable	Set-point
CL1	40	4	10	2	1000	Volume	12.15
CL2	40	4	10	2	0.1	Supersaturation	1.15
CL3	60	4	15	2	0.01	Supersaturation	1.15
CL4	80	4	20	2	10000	Fraction of crystals	0.43

5. PID CONTROLLERS

The PID parameters were tuned, where k_p, τ_i, τ_d are related with the general PID terminology as follows (Aström & Hägglund, 1995):

$$u(t+k) = K_p\left[e(t+k) + \frac{\Delta t}{\tau_i} \cdot \sum_{i=0}^{k} e(t+i) + \frac{\tau_d}{\Delta t} \cdot \left(e(t+k) - e(t+k-1)\right)\right] \tag{11}$$

Since the process is nonlinear, classical (linear) tuning procedures were substituted by a numerical optimization of the integral (or sum in the discrete version) of the absolute error (IAE):

$$IAE = \sum_{k=1}^{N} \left|ref(t+k) - y_p(t+k)\right| \tag{12}$$

Equation (12) was minimized in a closed loop framework between the discrete process model and the PID controller. For each parameter an interval of possible values was defined based on empirical knowledge and the process operator expertise. A number of gradient (Newton-like) optimization methods were employed to compute the final values of each controllers summarized in Table 1. All methods concluded that the derivative part of the controller is not necessary. Hence, PI controllers were analyzed in the next tests.

Table 3. Optimized PID parameters for the control loops define in Table 1

	Control loop 1	Control loop 2	Control loop 3	Control loop 4
k_p	0.05	-0.5	20	-0.01
τ_i	30	40	10	70
τ_d	0	0	0	0

6. DISCUSSION OF RESULTS

The operation strategy, summarized in Table 1 and implemented by a sequence of Classical- MPC, NNMPC or PI controllers is comparatively tested in Matlab environment. The output predictions are provided either by a simplified discrete model (with the main operation parameters kept constant) or by a trained ANN model (5-8). A process simulator was developed based on a detailed phenomenological model (Georgieva et al., 2003). Realistic disturbances and noise are introduced substituting the analytical expressions for the vacuum pressure, brix and temperature of the feed flow, pressure and temperature of the steam with original industrial data (without any preprocessing(Scenario-2)). The test is implemented for two different scenarios of work.

- **Scenario - 1:** The simulation uses, like process, the set of equations differentials proposed in (Georgieva et al. 2003) with empirical operation parameters.
- **Scenario - 2:** The simulation uses, like process, the set of equations differentials proposed in (Georgieva et al. 2003), but are used like operation parameter e real industrial data batch not used in neural network training.

Time trajectories of the controlled and the manipulated variables for the control loop 1, 2 and 4 of one batch (Batch 1) are depicted in Figs. 4-6. The three controllers guarantee good set point tracking. However, the quality of the produced sugar is evaluated only at the process end by the crystal size distribution (CSD) parameters, namely AM and CV. The results are summarized in Table 4 and both classical and NNPMC outperform the PI. Our general conclusion is that the main benefits of the MPC strategy are with respect to the batch end point performance.

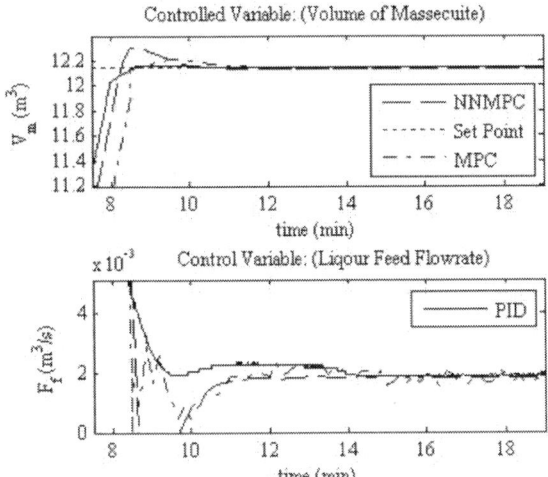

Figure 4. Controlled (Volume of massecuite) and control variables (F_f- feed flowrate) over time for the 1st control loop.

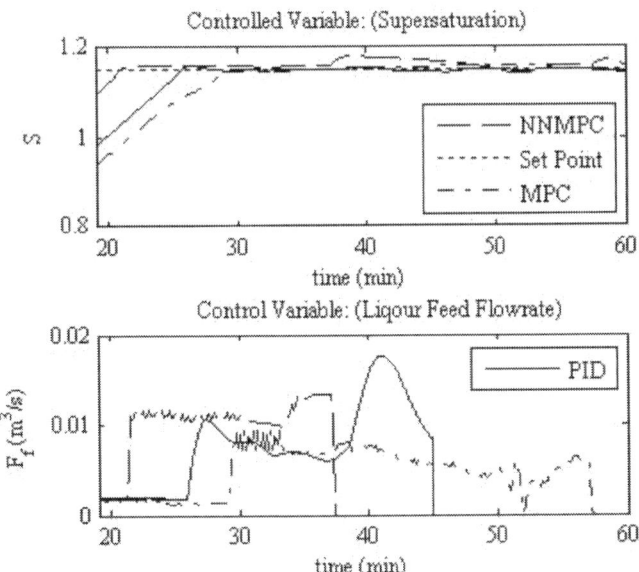

Figure 5. Controlled (Supersaturation) and control variables (F_f- feed flowrate) over time for the 2nd control loop.

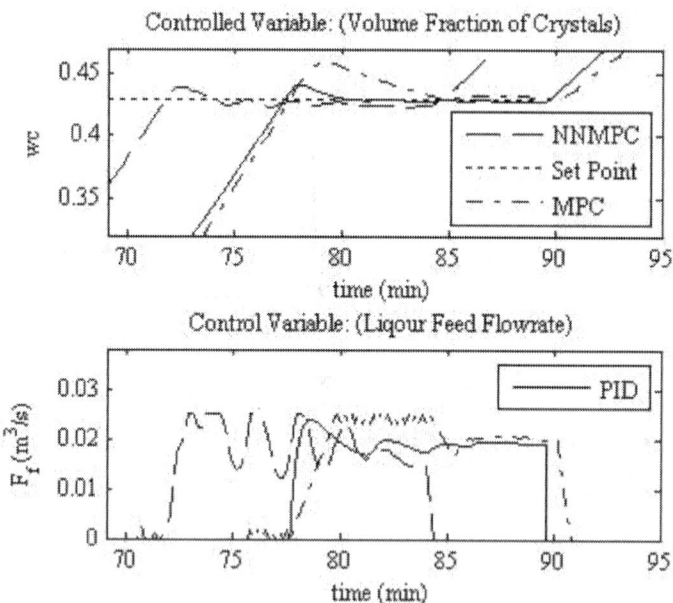

Figure 6. Controlled (Volume fraction of crystals) and control variables (F_f-feed flowrate) over time for the 4th control loop.

Table 4. Batch end point performance measures (Batch - 1)

Performance measures	Classical MPC	NN-MPC	PI
AM (mm) (reference 0.56)	0.586	0.584	0.590
CV (%)	32.17	31.13	32.96

Table 5. Batch end point performance measures (Batch - 2)

Performance measures	Classical MPC	NN-MPC	PI
AM (mm) (reference 0.56)	0.615	0.609	0.613
CV (%)	29.39	30.28	31.14

Table 6. Batch end point performance measures (Batch - 3)

Performance measures	Classical MPC	NN-MPC	PI
AM (mm) (reference 0.56)	0.636	0.631	0.639
CV (%)	28.74	29.42	29.23

7. CONCLUSION

With the results obtained in this work it has been demonstrated that algorithm NNMPC is a viable solution to control nonlinear complexes processes, still in the case that only exists input-output information of the process.

An aspect very important to obtain successful results with NNMPC is the representative quality of the available data, which was demonstrated with the results obtained in the third control loop analyzed.

The weighting factor λ_2 has a crucial paper in the good NNMPC performance. A constrain very hard can impose that the control signal can not follow the dynamics of the process, but a very soft constrain can cause instability in the control signal, when the model is not precise.

8. ACKNOWLEDGEMENTS

Several institutions contributed for this study: 1) Foundation of Science and Technology of Portugal, which financed the scholarship of investigation of doctorate SFR/16175/2004; 2) Laboratory for Process, Environmental and Energy Engineering (LEPAE), Department of Chemical Engineering, University of Porto; 3) The Institute of Electronic Engineering and Telematics of Aveiro (IEETA); 4) Sugar refinery RAR, Portugal; The authors are thankful to all of them.

9. APPENDIX A. CRYSTALLIZATION MODEL

Sugar crystallization occurs through the mechanisms of nucleation, growth and agglomeration. The general phenomenological model of the fed-batch crystallization process consists of mass, energy and population balances, including the relevant kinetic rates for nucleation, linear growth and agglomeration [Ilchmann, et al., 1994]. While the mass and energy balances are common expressions in many chemical process models, the population balance is related with the crystallization phenomenon, which is still an open modeling problem.

Mass balance
The mass of all participating solid and dissolved substances are included in a set of conservation mass balance equations:

$$\dot{M} = f_1(M(t), F(t), S_1(t)), \quad t_0 \leq t \leq t_f, \quad M(0) = M_0$$

(A-1)

where $M(t) \in \Re^q$ and $F(t) \in \Re^m$ are the mass and the flow rate vectors, with q and m dimensions respectively, and t_f is the final batch time. $S_1(t) \in \Re^{r1}$ is the vector of physical time dependent parameters as density, viscosity, purity, etc. For the process in hand, the detailed form of the macro-model (A1) is as follows

$$M_{sol} = M_a + M_i + M_w \tag{A-2}$$

$$M_m = M_{sol} + M_c \tag{A-3}$$

$$\frac{dM_w}{dt} = F_f \rho_f \left(1 - B_f\right) + F_w \rho_w - J_{vap} \tag{A-4}$$

$$\frac{dM_i}{dt} = F_f \cdot \rho_f \cdot B_f \cdot \left(1 - Pur_f\right) \tag{A-5}$$

$$\frac{dM_a}{dt} = F_f \cdot \rho_f \cdot B_f \cdot Pur_f - J_{cris} \tag{A-6}$$

$$\frac{dM_c}{dt} = J_{cris} \tag{A-7}$$

$$V_m = \frac{M_c + M_{sol}}{\rho_{sol}} \tag{A-8}$$

$$J_{vap} = \frac{W + Q}{\lambda_{vap}} + K_{vap} \cdot \left(T_m - T_{w(vac)} - BPE\right) \tag{A-9}$$

Energy balance

The general energy balance model is

$$\frac{dT_m}{dt} = aJ_{cris} + bF_f + cJ_{vap} + d \tag{A-10}$$

where parameters a, b, c and d incorporate the enthalpy terms and specific heat capacities derived as time dependent functions of physical and thermodynamic properties as follows

$$a = \frac{H_{sol} - H_c + (1 - B_{sol})\dfrac{dH_{sol}}{dB_{sol}} + \dfrac{1 - Pur_{sol}}{B_{sol}} \cdot \dfrac{dH_{sol}}{dPur_{sol}}}{M_{sol} \cdot Cp_{sol} + M_c \cdot Cp_c}$$

(A-11)

$$b = \frac{\rho_f \left(H_f - H_{sol} + (B_f - B_{sol})\dfrac{dH_{sol}}{dB_{sol}} + \dfrac{B_f\left(Pur_f - Pur_{sol}\right)}{B_{sol}} \cdot \dfrac{dH_{sol}}{dPur_{sol}} \right)}{M_{sol} \cdot Cp_{sol} + M_c \cdot Cp_c}$$

(A-12)

$$c = \frac{H_{sol} - H_{vap} - B_{sol} \cdot \dfrac{dH_{sol}}{dB_{sol}}}{M_{sol} \cdot Cp_{sol} + M_c \cdot Cp_c}$$

(A-13)

$$d = \frac{W + Q + F_w \rho_w (H_w - H_{sol} + B_{sol})\dfrac{dH_{sol}}{dB_{sol}}}{M_{sol} \cdot Cp_{sol} + M_c \cdot Cp_c}$$

(A-14)

$$\frac{dH_{sol}}{dB_{sol}} = -29.7 T_m + 4.6 Pur_{sol} T_m + 0.075 T_m^2$$

(A-15)

$$\frac{dH_{sol}}{dPur_{sol}} = 4.61 Bx_{sol} T_m$$

(A-16)

Population balance

Mathematical representation of the crystallization rate can be achieved through basic mass transfer considerations or by writing a population balance represented by its moment equations. Employing a population balance is generally preferred since it allows to take into account initial experimental distributions and, most significantly, to consider complex mechanisms such as those of size dispersion and/or particle agglomeration/aggregation. The basic moments of the number-volume distribution function are

$$\frac{d\tilde{\mu}_0}{dt} = \tilde{B}_0 - \frac{1}{2} \cdot \beta' \cdot \tilde{\mu}_0^2$$

(A-17)

$$\frac{d\tilde{\mu}_1}{dt} = G_v \cdot \tilde{\mu}_0$$

(A3-18)

$$\frac{d\tilde{\mu}_2}{dt} = 2 \cdot G_v \cdot \tilde{\mu}_1 + \beta' \cdot \tilde{\mu}_1^2$$

(A3-19)

$$\frac{d\tilde{\mu}_3}{dt} = 3 \cdot G_v \cdot \tilde{\mu}_2 + 3 \cdot \beta' \cdot \tilde{\mu}_2^2$$

(A3-20)

$$J_{cris} = \rho_c \cdot \frac{d\tilde{\mu}_1}{dt},$$

(A3-21)

where $\tilde{B}_0$, G and β' are the kinetic variables nucleation rate, linear growth rate and the agglomeration kernel, respectively with the following mathematical descriptions

$$\tilde{B}_0 = K_n \cdot 2.894 \cdot 10^{12} \cdot G^{0.51} \cdot \left(\frac{\tilde{\mu}_1}{k_v \cdot V_m} \right)^{0.53} \cdot V_m$$

(A-22)

$$\beta' = \frac{K_{ag} \cdot G \cdot \tilde{\mu}_1}{V_m^2}$$

(A3-23)

$$G = K_g \cdot \exp\left(-\frac{57000}{R(T_m + 273)} \right) \cdot (S-1) \cdot \exp(-13.863(1 - P_{sol})) \cdot \left(1 + 2 \cdot \frac{v}{V_m} \right)$$

(A-24)

$$G_v = 3 \cdot k_v \left(\frac{v}{\tilde{\mu}_0} \right)^{2/3} \cdot G.$$

(A-25)

The crystallization quality is evaluated by the particle size distribution (PSD) at the end of the process which is quantified by two parameters - the final average (in mass) particle size (AM) and the final coefficient of particle variation (CV) with the following definitions:

$$AM = \bar{L} \tag{A-26}$$

$$CV = \frac{\sigma}{\bar{L}} \tag{A-28}$$

Where σ and $\bar{L}$ are computed from:

$$\bar{L} = \left(\frac{\eta_3}{1 + 3 \cdot \left(\frac{\sigma}{\bar{L}}\right)^2} \right)^{1/3} \tag{A-29}$$

$$15 \cdot \eta_3^2 \cdot \left(\frac{\sigma}{\bar{L}}\right)^6 + \left(45 \cdot \eta_3^2 - 9 \cdot \eta_6\right)\left(\frac{\sigma}{\bar{L}}\right)^4 + \left(15 \cdot \eta_3^2 - 6 \cdot \eta_6\right)\left(\frac{\sigma}{\bar{L}}\right)^2 + \eta_3^2 - \eta_6 = 0 \tag{A-30}$$

In (A-29, A-30), η_j represent moments of mass-size distribution functions, that are related to the moments of the number-volume distribution functions μ_j by the following relationships:

$$\eta_3 = \frac{\mu_2}{k_v \cdot \mu_1}, \tag{31}$$

and

$$\eta_6 = \frac{\mu_3}{k_v^2 \cdot \mu_1} \tag{A3-32}$$

Correlations for physical properties

$$Q = \alpha_s \cdot F_s \cdot \Delta H_s \tag{A-33}$$

$$\rho_f = \left(1000 + \frac{Bx_f \cdot (200 + Bx_f)}{54} \right) \cdot \left(1 - 0.036 \cdot \frac{T_f - 20}{160 - T_f} \right) \tag{A-34}$$

$$Cp_f = 4186.8 - 29.7 \cdot Bx_f + 4.61 \cdot Bx_f \cdot Pur_f + 0.075 \cdot Bx_f \cdot T_f \tag{A-35}$$

$$H_f = Cp_f \cdot T_f \tag{A-36}$$

$$\rho^*_{sol} = \left(1000 + \frac{Bx_{sol} \cdot (200 + Bx_{sol})}{54}\right) \cdot \left(1 - 0.036 \cdot \frac{T_m - 20}{160 - T_m}\right) \tag{A-37}$$

$$\rho_{sol} = \rho^*_{sol} + 1000 \cdot \left(-1 + \exp\left[\left(-6.927 \cdot 10^{-6} \cdot Bx_{sol}^2 - 1.164 \cdot 10^{-4} \cdot Bx_{sol}\right) \cdot (Pur_{sol} - 1)\right]\right) \tag{A-38}$$

$$Cp_{sol} = 4186.8 - 29.7 \cdot Bx_{sol} + 4.61 \cdot Bx_{sol} \cdot Pur_{sol} + 0.075 \cdot Bx_{sol} \cdot T_m \tag{A-39}$$

$$H_{sol} = Cp_{sol} \cdot T_m \tag{A-40}$$

$$\rho_m = \frac{\rho_{sol} \cdot \rho_c}{\rho_c - w_c \cdot (\rho_c - \rho_{sol})} \tag{A-41}$$

$$Pur_{sol} = \frac{M_a}{M_a + M_i} \tag{A-42}$$

$$B_{sol} = \frac{M_a + M_i}{M_{sol}} \tag{A-43}$$

$$Bx_{sol} = 100 \cdot B_{sol} \tag{A-44}$$

$$Bx_{sat} = 64.447 + 8.222 \cdot 10^{-2} \cdot T_m + 1.66169 \cdot 10^{-3} \cdot T_m^2 - 1.558 \cdot 10^{-6} \cdot T_m^3 - 4.63 \cdot 10^{-8} \cdot T_m^4 \tag{A-45}$$

$$S^* = 1.129 - 0.284 \cdot (1 - Pur_{sol}) + (2.333 - 0.0709 \cdot (T_m - 60)) \cdot (1 - Pur_{sol})^2 \tag{A-46}$$

$$S = \frac{\dfrac{Bx_{sol}}{100 - Bx_{sol}}}{\dfrac{Bx_{sat}}{100 - Bx_{sat}} \cdot C_{sat}} \tag{A-47}$$

$$C_{sat} = 0.1 \cdot \frac{Bx_{sol}}{100 - Bx_{sol}} \cdot (1 - Pur_{sol}) + 0.4 + 0.6 \cdot \exp\left(-0.24 \cdot \frac{Bx_{sol}}{100 - Bx_{sol}} \cdot (1 - Pur_{sol})\right) \tag{A-48}$$

$$v = \frac{M_c}{\rho_c} \tag{A-49}$$

$$w_c = \frac{M_c}{M_c + M_{sol}}$$

(A-50)

$$Cp_c = 1163.2 + 3.488 \cdot T_m$$

(A-51)

$$H_c = Cp_c \cdot T_w$$

(A-52)

$$\rho_w = 1016.7 - 0.57 \cdot T_w$$

(A-53)

$$T_{w(vac)} = 122.551 \cdot \exp\left(-0.246 \cdot P_{vac}\right) \cdot \left(P_{vac}\right)^{0.413}$$

(A-54)

$$T_{w(s)} = 100.884 \cdot \exp\left(-1.203 \cdot 10^{-2} \cdot P_s\right) \cdot \left(P_s\right)^{0.288}$$

(A-55)

$$\lambda_{w(vac)} = 2263.28 - 58.21 \cdot \ln\left(P_{vac}\right)$$

(A-56)

$$\lambda_s = 2257.51 - 85.95 \cdot \ln\left(P_s\right)$$

(A-57)

$$H_w = 2323.3 + 4106.7 \cdot T_w + T_w^{\,2}$$

(A-58)

$$H_{w(s)} = 2323.3 + 4106.7 \cdot T_{w(s)} + T_{w(s)}^{\,2}$$

(A-59)

$$H_s = 2491860 - 13270 \cdot P_s + \left(1946.5 + 37.9 \cdot P_s\right) \cdot T_s$$

(A-60)

$$H_{vac} = 2499980 - 24186 \cdot P_{vac} + \left(1891.1 + 106.1 \cdot P_{vac}\right) \cdot T_m$$

(A-61)

$$\Delta H_s = H_s + H_{w(s)}$$

(A-62)

$$BPE = \left(0.03 - 0.018 \cdot Pur_{sol}\right) \cdot \left(T_{w(vac)} + 84\right) \cdot \left(\frac{Bx_{sol}}{100 - Bx_{sol}}\right)$$

(A-63)

For more detailed presentation of the process model, refer to [Georgieva et al., 2003].

REFERENCES

1. F. Allgöwer, R. Findeisen, Z. K. Nagy, 2004 Nonlinear model predicitve control: From theory to application. Journal of Chinese Institute of Chemical Engineers, 35 (3), 299-315.
2. K. J. Aström, T. Hägglund, 1995 Pid controllers : theory, design, and tuning. North Carolina: Research Triangle Park, Instrument Society of America.
3. L. S. Balasubramhanya, F. J. Doyle, 2000 Nonlinear model-based control of a batch reactive distillation column. Journal of Process Control, 10 209218 .
4. Bemporad, M. Morari, N. L. Ricker, 2005 User's Guide: Model predictive control toolbox for use with MatLab: The MathWorks Inc.
5. E. F. Camacho, C. Bordons, 2004 Model predictive control in the process industry. London: Springer-Verlag.
6. J. M. Chorão, 1995. Operação assistida por comutador dum cristalizador industrial de açúcar, Ph. D. Tesis, Faculdade de Engenharia, Departamento de Eng. Química, Universidade de Porto, Porto
7. M. Diehl, H. G. Booc, J. P. Schlder, R. Findeisen, A. Nagy, F. Allgöwer, 2002 Real-time optimization and nonlinear model predictive control of processes governed by deferential algebraic equations. Jornal of Process Control 12 577585 .
8. Azevedo. S. Feyo de, M. J. Gonçalves, 1988 Dynamic Modelling of a Batch Evaporative Crystallizer. Recent Progrés en Génie de Procedés, Lavoisier, Paris: Ed. S. Domenech, X. Joulia, B. Koehnet, 199204 .
9. P. Georgieva, M. J. Meireles, Azevedo. S. Feyo de, 2003 Knowledge Based Hybrid Modeling of a Batch Crystallization When Accounting for Nucleation, Growth and Agglomeration Phenomena. Chemical Engineering Science, 58 36993707 .
10. S. J. Jancic, P. A. M. Grootscholten, 1984 Industrial Crystallization. Delft, Holland: Delft University Press.
11. M. Morari, 1994 Advances in Model-Based Predictive Control. Oxford: Oxford University Press.
12. S. J. Qin, T. A. Badgwell, 2003 A survey of model predictive control technology. Control Engineering Practice 11 7 733764 .
13. J. Rawlings, 2000 Tutorial Overview of Model Predictive Control. IEEE Control Systems Magazine:3852 .
14. J. A. Rossiter, 2003 Model based predictive control. A practical approach. New York: CRC Press.
15. H. Seki, M. Ogawa, S. Ooyama, K. Akamatsu, M. Ohshima, W. Yang, 2001 Industrial application of a nonlinear model predictive control to polymerization reactors. Control Engineering Practice, 9 819828 .

16. Simoglou, P. Georgieva, E. B. Martin, J. Morris, Azevedo. S. Feyo de, 2005 On-line Monitoring of a Sugar Crystallization Process. Computers & Chemical Engineering, 29 (6), 14111422 .

17. R. Soeterboek, 1992 Predictive control. A unified approach. New York: Prentice Hall International.

CHAPTER 4

Predictive Control for Active Model and its Applications on Unmanned Helicopters

Dalei Song1,Juntong Qi, Jianda Han And Guangjun Liu

[1] Shenyang Institute of Automation Science, Chinese Academy of Sciences, China

1. INTRODUCTION

Unmanned helicopters are increasingly popular platforms for unmanned aerial vehicles (UAVs). With the abilities such as hovering, taking off and landing vertically, unmanned helicopters extend the potential applications of UAVs. However, due to the complex mechanism and complicated aero-flow during flight, it is almost impossible to accurately model the dynamics of an unmanned helicopter in full flight envelope, and the significant model uncertainties associated with a nominal model may degrade the performance and even stability of an onboard controller.

Due to the difficulty in obtaining a high fidelity full envelope model, the multi-mode modeling technique has been proposed for rotor aircrafts, such as tilt-rotor aircraft XV-15 [1], helicopter BO-105 [2], UH-60 [3], R-50 [4] and X-Cell [5]. The mode-dependent model, which is identified and simplified according to a specific flight mode, such as hovering, cruising, taking off and landing, can be used for control design for the corresponding flight mode. However, the mode-dependent control suffers from at least two problems: one is the difficulty in accommodating the mode transition dynamics, and the other is the compensation of the 'model drift' due to flight dynamics change within one particular mode. Up to now, for the purpose of practical implementation, the mode transition problem can be partially dealt with by limiting the mode switching conditions [6], e.g., mode change is made through hovering mode.

Robust and adaptive control techniques [7-8], on the other hand, have been used to deal with the 'model-shift' within a flight mode. However, such control schemes normally need to know the boundary of internal and external uncertainties and relative noise distribution, which are difficult to identify accurately for a helicopter in full flight envelope. Although online identification technology can be used to obtain the real-time dynamics and disturbance, it is a large burden for the flight computer to reconstruct the robust controllers and reach the requested control period (>50Hz) for sampling and actuating due to the complex calculation of the robust/adaptive optimization process [9-10] and the strict weight limits of micro flight computers.

Besides the model uncertainties, another critical problem that limits the control performance of a helicopter is the time delay between the actuator command and the generation of relative aerodynamic force/torque [11], which will be called aerodynamics-delay/time-delay in the following sections. Normally, this time delay may cause reduced feedback gain of a model-based controller and result in poor robustness [12-13], i.e., sensitive to disturbances.

In recent years, the encouraging achievement in sequential estimation makes it an important direction for online modeling and model-reference control [14]. Among stochastic estimations, the most popular one is the Kalman-type filters (KFs) [15, 16, and 17]. Although widely used, the KFs suffer from sensitivity to bias and divergence in the estimates, relying on assumptions on statistic distribution such as white noise and known mean or covariance for optimal estimation. In many cases, it is more practical to assume that the noises or uncertainties are unknown but bounded (UBB). In view of this, the set-membership filter (SMF), which computes a compact feasible set in which the true state or parameter lies only under the UBB noise assumption, provides an attractive alternative [18-19].

On the control issue, model predictive control (MPC) can compensate for the aerodynamics delay and does not require a high accuracy reference nonlinear model [20]. Among these methods, linear generalized predictive control (GPC) has become one of the most popular MPC methods in industry and academia. However, the normal GPC is sensitive to process noise and model errors [21], which are unknown but bounded for helicopters when sudden 'mode change' happen and model-drift in full flight envelope. This makes the prediction biased, and results in the non-optimal process of controller solving.

In this paper, for realizing the coupling control of unmanned helicopters in full flight envelope, an active modeling based controller is developed based on a modified generalized predictive control and adaptive set-membership filter estimation (ASMF). The time varying model error and its boundary are estimated by the adaptive set-member filter, which is first proposed in [19]. Incremental prediction process and dimension reduction method is embedded into traditional GPC, which can decreases the computation burden and maintain prediction unbiased when 'mode change' happens. Based on this active estimation and the modified GPC controller, a novel optimal strategy for on-line compensation of model error is developed. Thus, aggressive flight can be achieved only based on the hovering model with time-delay terms. Using the identified hovering dynamics model as nominal model for controller, flight experiments have been conducted to test the performance of the proposed controller in full flight envelope on our UAV platform, and experimental results have demonstrated the effectiveness of the proposed method.

2. ACTIVE MODEL BASED CONTROL SCHEME AND REFERENCE MODEL OF A HELICOPTER

Fig. 1 illustrates the active model based control scheme. The error between the reference model and the actual dynamics of the controlled plant is estimated by an on-line modeling strategy. The control, which is designed according to the reference model, should be able to compensate the estimated model error and it in real time. In the followings of this paper, we use the ASMF as the active modeling algorithm and the modified GPC as the control.

For normal missions of an unmanned helicopter, the flight modes include hovering (velocity under 5m/s), cruising (velocity above 5m/s), taking off and landing (distance to the ground is below 3m while significant ground effect exists) and the transitions among these modes. A reference model is typically obtained by linearizing the nonlinear dynamics of a helicopter at one flying mode. The model errors from linearization, external disturbance, simplification, and un-modeled dynamics can be considered as additional process noise [22]. Thus, a linearized state-space model for helicopter dynamics in full flight envelope can be formulated as

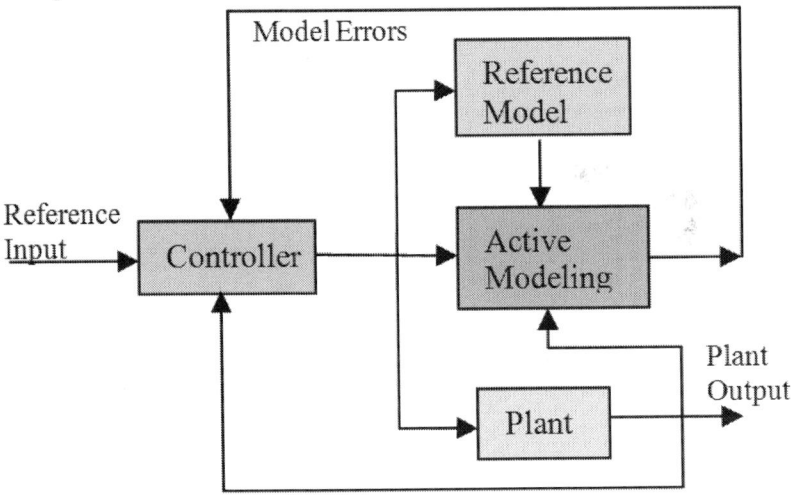

Figure 1. The scheme of active model based control

$$\begin{cases} \dot{X}_t = A_0 X_t + B_0 U_{t-k} + B_f f(X_t, \dot{X}_t, W_t) \\ Y_t = C X_t \end{cases}$$

(1)

where $X \in R^{13}$ is the state, including 3-axis velocity, pitch and roll angle, 3-axis angle rate, flapping angles of main rotor and stabilizer bar, and the feedback of yaw gyro. $Y_t \in R^8$ is the output, including 3-axis velocity, pitch and roll angle and 3-axis angle rate, A_0 and B_0 contain parameters that can be identified in different flight modes, and we use them to describe the parameters in hovering mode. $U \in R^4$ is the control input vector. $C \in R^{13 \times 8}$ is the output matrix, $k \in R$ is the time-delay for the driving system. The detail of building the nominal model and physical meaning of parameters is explanted in Appendix A.

To describe the dynamics change, in equation (1), here, we introduce $f(X_t, \dot{X}_t, W_t) \in R^{13}$ to represent the time varying model error in full flight envelope, and $W_t \in R^{13}$ is the process noise.

The following two sections, based on model (1) will describe the way to estimate $f(X_t, \dot{X}_t, W_t)$ and to compensate for model errors from process noise, parameters change, control delay and flight mode change in real applications.

3. ASMF BASED ACTIVE MODEL ERROR ESTIMATION

As illustrated in Fig.1, adopting the active modeling process to get the model error f and system state X is the basis for elimination of the model error. Controller can only work based on nominal model and feedback of state and model error from active modeling process. In this section, the active modeling process is built based on an adaptive set-membership filter (ASMF) [19] since the UBB process noise.

First, we must obtain the reference equation for estimation. Compared with the sampling frequency (often >50Hz for flight control) of the control system, the model error $f(X, \dot{X}, W)$ can be considered as a slow-varying vector, which means

$$f_{t+1} = f_t + h_t$$

where f_t is the sampling value of $f(X, \dot{X}, W)$ at sampling time t, and h_t is the assumed unknown but bounded (UBB) process noise.

Let the extended sampling state

$$X_t^a = \left(X_t^T \quad f_t^T \right)^T$$

Then, we can obtain the discrete equation from Eq. (1) as

$$\begin{cases} X_{t+1}^a = A_d^a X_t^a + B_d^a U_t + W_t^a \\ Y_t = C_d^a X_t^a + V_t \end{cases}$$

$$(2)$$

where
$$A_d^a = \begin{pmatrix} A_d & B_f \\ 0_{13\times13} & I_{13\times13} \end{pmatrix}, \quad B_d^a = \begin{pmatrix} B_d \\ 0_{13\times4} \end{pmatrix}, \quad C_d^a = (C_d \quad 0_{8\times13}), \quad W_t^a = (W_t^T \quad h_t^T)^T, \quad B_f = I_{13\times13}$$

and f_t is a 13×1 vector for model errors. Here, t is the sampling time, $I_{m\times m}$ is the $m\times m$ unit matrix and $0m\times n$ is the $m\times n$ zero matrix. $\{A_d, B_d, C_d\}$ is the discrete expression of system $\{A_0, B_0, C\}$. Here, time-delay k is ignored during the estimate process, and the compensation method will be discussed in the next part on modified GPC.

The model error i.e., f in Eq. (1), comes from the linearization while neglecting the coupling dynamics and uncertainties, and also the A_0 and B_0 because they are identified with respect to a specific flight mode, here hovering mode is selected as nominal flight mode since easy identification. Therefore, both the model error and the process noise W^a are vehicle dynamics and flight states dependent, and do the following assumption

Assumption:
W^a does not necessarily have a normal distribution.

Thus, the Kalman type filter cannot be applied, and adaptive set-membership filter, which is developed for UUB process noise and can get the uncertain boundaries of the states, is considered to estimate the states and model errors here.

In this section we only present the result of ASMF and please refer to [19] for the details about ASMF. With respect to Eq. (2), we can build the adaptive set-membership filter as Eq. (3), where Q^a and R^a are the initial elliptical boundary of process and measurement noise respectively, rm is the maximum eigenvalue of R, p_m is the maximum eigenvalue of $C_d^a P_{t|t-1} C_d^{aT}$, $Tr(\bullet)$ is the trace of a matrix, δ_t and β_t are the adaptive parameters of the filter. We can also obtain the boundary of the ith element $\hat{X}_i^a$ of extended state $\hat{X}_{t|t}^a$ as $\left(\hat{X}_i^a - \sqrt{P_{ii}}, \hat{X}_i^a + \sqrt{P_{ii}}\right)$, where P_{ii} is the i-th diagonal element of matrix P.

$$\begin{cases} \rho_t = \dfrac{\sqrt{r_{mt}}}{\sqrt{r_{mt}} + \sqrt{p_{mt}}} \\[3mm] W_t = \dfrac{C_d^a P_{t|t-1} C_d^{aT}}{1-\rho_t} + \dfrac{R^a}{\rho_t} \\[3mm] K_t^e = \dfrac{P_{t|t-1} C_d^{aT} W_t^{-1}}{1-\rho_t} \\[3mm] \delta_t = 1 - (Y_t - C_d^a \hat{X}_{t|t-1}^a)^T W_t^{-1}(Y_t - C_d^a \hat{X}_{t|t-1}^a) \\[3mm] \hat{X}_{t|t}^a = \hat{X}_{t|t-1}^a + K_t^e(Y_t - C_d^a \hat{X}_{t|t-1}^a) \\[3mm] P_{t|t} = \delta_t \left(\dfrac{P_{t|t-1}}{1-\rho_t} - \dfrac{P_{t|t-1}}{1-\rho_t} C_d^{aT} W_t^{-1} C_d^a \dfrac{P_{t|t-1}}{1-\rho_t} \right) \\[3mm] \hat{X}_{t+1|t}^a = A_d^a \hat{X}_{t|t}^a + B_d^a U_t \\[3mm] \beta_t = \dfrac{\sqrt{Tr(Q^a)}}{\sqrt{Tr(Q^a)} + \sqrt{Tr(A_d^a P_{t|t} A_d^{aT})}} \\[3mm] P_{t+1|t} - \dfrac{A_d^a P_{t|t} A_d^{aT}}{1-\beta_t} + \dfrac{Q^a}{\beta_t} \end{cases} \qquad (3)$$

4. MODIFIED GPC FOR UNMANNED HELICOPTERS

To eliminate the negative influence of model errors and control delay in flight, besides the active estimation algorithm like ASMF that does not require a normal distribution assumption, an effective control algorithm has to be designed according to the reference model of Eq. (1) while adopting the on-line estimation of f as compensation.

We describe the normal GPC in Section 4.1, and then, the modified scheme is proposed in Section 4.2 & 4.3 to eliminate the negative influence of model errors in real applications.

4.1. Preliminary Work For Generalized Predictive Control

Generally, for a linear system with actuator time delay like,

$$\begin{cases} X_{t+1} = A_d X_t + B_d u_{t-k} + W_t \\ y_t = C_d X_t \end{cases} \tag{4}$$

where $X_t \in R^{n \times 1}$ is the system state vector at sampling time t, $y_t \in R^{l \times 1}$ is the output vector, $u_t \in R^{m \times 1}$ is the control input vector, k is the actuators' time-delay and W_t is process noise; traditional Generalized Predictive Control (GPC) [23] can be designed as: Step I: Make prediction

Firstly, for the case that predictive step i is less than time-delay k (i.e., the time instant that system behavior cannot be regulated through current and future control action), prediction can be denoted as following equation,

$$\hat{X}_{t+i|t} = A_d \hat{X}_{t+i-1|t} + B_d u_{t+i-1-k} \triangleq \hat{X}^1_{t+i|t} \tag{5}$$

where $\hat{X}_{t+i}|t$ is the prediction state at time $t+i$, the superscript 1 denotes that the part of predicted variable that is independent of the current and future's control actions.

Secondly, for the case that prediction step i is larger than the time delay k,

$$\hat{X}_{t+k+i|t} = A_d \hat{X}_{t+k+i-1|t} + B_d u_{t+i}$$

$$= A_d \hat{X}^1_{t+k+i-1|t} + \sum_{n=0}^{i} A_d^n B_d u_{t+n}$$

$$= A_d \hat{X}^1_{t+k+i|t} + \sum_{n=0}^{i} A_d^n B_d u_{t+n} \quad , \quad 1 \le i \le p \tag{6}$$

where p is the prediction range; similarly, $\hat{X}^1_{t+k+i-1|t}$ denotes the sub-variable of $\hat{X}_{t+k+i-1|t}$ that is independent of the current and future's control actions.

Step II: Receding horizon optimization

After making prediction, the control vector can be obtained by minimize the following cost function:

$$J = (R^x_t - X^p_t)^T (R^x_t - X^p_t) + U_t^T \gamma U_t \tag{7}$$

And the optimal control inputs can be denoted as,

$$U_t^* = (G_0^T G_0 + \gamma)^{-1} G_0^T (R_t^x - X_t^1)$$

(8)

where G_0 is the predictive matrix, X_t^v is the predictive state vector, X_t^1 is the known vector inside X_t^v, λ is the weight of control input, and R_t^x is the reference of system states. The detailed definition of these matrixes can be referenced in [23].

Step III: Control implementation

The first element of vector U_t^* is used as the control to the real plant. After that, go back to step I at the next time instant.

However, with application to the unmanned helicopters, this kind of GPC algorithm has the following three disadvantages, which will be solved in the next two sections:

1. It cannot reject the influence of working mode changes, i.e., if

$$X_t = x_t - x_0 > \pi(x_0, u_0)$$
$$U_t = u_t - u_0 > \pi(x_0, u_0)$$

(9)

where (x_0, u_0) is the current operation point, which cannot be ensured on-line, $\pi(x_0, u_0)$ is the valid range for model linearization and x_t is the absolute state at time t, u_t is the absolute control input at time t. The biased prediction, due to the changing operation point (x_0, u_0), will bring steady errors for velocity tracking.

2. Normal GPC is sensitive to mismatch of the nominal model, which means slow change in parameters (A_d, B_d) may result in prediction error and unstable control.

3. The transient model errors of the nominal model from external disturbance, estimated by ASMF, cannot be eliminated. And this will also result in the non-minimum variance and the instability of the closed control loop.

4.2. Stationary Increment Predictive Control

To reject the influence of working mode change and sensitivity to nominal parameters change in real application, i.e. the problem 1) and 2) in Section 4.1, we assume that the process noise W_i's increment in Eq. (4) is a stationary random process, which means

$$W_t^0 \triangleq \Delta W_t = W_t - W_{t-1}$$

(10)

is normal distribution. Where $\Delta = 1 - q^{-1}$ is the difference operator; q^{-1} is one-step delay factor. Thus, Eq. (4) can be rewritten as follows,

$$\Delta X_{t+1} = A_d \Delta X_t + B_d \Delta u_{t-k} + W_t^0$$

(11)

Consider

$$\Delta X_t = (x_t - x_0) - (x_{t-1} - x_0) = \Delta x_t$$
$$\Delta U_t = (u_t - u_0) - (u_{t-1} - u_0) = \Delta u_t$$

if behavior prediction is made based on Eq. (11), only the absolute state x_t and control input u_t, which can be measured or estimated directly from sensors, are used and the current operation point (x_0, u_0) disappears in prediction. Thus, the problem of biased prediction due to changing of working point, i.e., problem 1), can be solved.

Otherwise, according to the process of traditional GPC, the set-point RxtRtx must be obtained for every prediction step, and this is often set as current reference states. However, for helicopter system, only measurable outputs are cared, such as position, velocity and etc; and the internal states, such as rotor's pitch angle and yaw gyro's feedback and so on, are coupled with the measurable states/outputs, and cannot be set independently. Others, this reference input often comes from position track planning, which changes quickly for flight and often cause a step-like signal for tracking. To avoid the step signal reference tracking, which is dangerous for unmanned helicopter system, we use a low pass filter to calculate the set-point inputs of the output in the future i-th step, $i=1, ..., p$.

Let $SP_t \in R^{l \times 1}$ results obtained in this work it has been

$$r_{t+k+i} = SP_t + \alpha(r_{t+k+i-1} - SP_t) \quad , 1 \le i \le p$$

(12)

where α is the cut-off frequency of the filter, the initial value $r_{t+k} = \hat{y}_{t+k|t}$, r_{t+k+i}, is the i-th set-point input, and $\hat{y}_{t+k|t}$ is the estimate of output at time $t+k$.

Thus, the set-point problem is solved and the output prediction can be implanted based on increment model (11) as follows:

When the prediction step i is less than time-delay k,

$$\hat{X}_{t+i|t} = \hat{X}_{t+i-1|t} + A_d \Delta \hat{X}_{t+i-1|t} + B_d \Delta u_{t+i-1-k} = \hat{X}^1_{t+i|t} \tag{13}$$

When the prediction step is larger than time-delay k, let

$$\Delta \hat{X}^1_{t+i|t} = \hat{X}^1_{t+i|t} - \hat{X}^1_{t+i-1|t}$$

Then,

$$\hat{X}_{t+k+i|t} = \hat{X}_{t+k+i-1|t} + A_d \Delta \hat{X}_{t+k+i-1|t} + B_d \Delta u_{t+i}$$

$$= \hat{X}^1_{t+k+i-1|t} + A_d \Delta \hat{X}^1_{t+k+i-1|t}$$

$$+ \sum_{m=0}^{i-1} \left\{ \{ \sum_{n=0}^{i-1-m} A_d^n B_d \} \Delta u_{t+m} \right\}$$

$$= \hat{X}^1_{t+k+i|t} + \sum_{m=0}^{i-1} \left\{ \{ \sum_{n=0}^{i-1-m} A_d^n B_d \} \Delta u_{t+m} \right\}, 1 \le i \le p \tag{14}$$

Hence, the above problem 1), which comes from working mode change, is solved because x_0 disappears in predictive equation (14).

We can obtain the following prediction matrix for the output, which is often cared in helicopter tracking problem, from Eq. (12) and (13):

$$\hat{Y}_t = \left(\hat{y}_{t+k+1|t} \quad \hat{y}_{t+k+2|t} \quad \cdots \quad \hat{y}_{t+k+p|t} \right)^T$$

$$= \left(C_d \hat{X}^1_{t+k+1|t} \quad C_d \hat{X}^1_{t+k+2|t} \quad \cdots \quad C_d \hat{X}^1_{t+k+p|t} \right)^T$$

$$+ G \left(\Delta u_t^T \quad \Delta u_{t+1}^T \quad \cdots \quad \Delta u_{t+p-1}^T \right)^T$$

$$= Y^1_t + G \Delta U \tag{15}$$

where Y^1_t is the known part of p steps' prediction, which cannot be influenced by current control input, and matrix G has the following form:

$$G = \begin{pmatrix} C_d B_d & 0 & \cdots & 0 \\ C_d B_d + C_d A_d B_d & C_d B_d & \cdots & 0 \\ \cdots & \cdots & \cdots & \cdots \\ C_d \sum_{i=0}^{p-1} A_d^i B_d & C_d \sum_{i=0}^{p-2} A_d^i B_d & \cdots & C_d B_d \end{pmatrix}$$

$$(16)$$

Compared with the normal GPC, the prediction of SIPC has better characteristics that can be described by the following theorem, which solves the above problem 2) in Section IV.A.

Theorem: for nominal model (11), when the nominal model parameters (Ad, Bd) change into (Adr, Bdr).

    1.   ∃$M, N > 0 \in R$, let the matrix norms satisfy

$$\|A_d\| < M, \|B_d\| < M$$

$$\|A_{dr}\| < N, \|B_{dr}\| < N$$

    2.   Define

$R_{max}\{\bullet\}$ is the operator for the maximum of eigenvalue of matrix $\bullet$.

Thus, if

$$R_{max}\{A_{dr}\} < 0$$

$$R_{max}\{A_d\} < 0$$

Then, the state prediction obtained by Eqs. (13-14) maintains unbiased, and the characteristic is also guaranteed in traditional GPC conditions, i.e. Eq. (4), where W_t is normal distribution.

Proof: See Appendix B.

In Eq. (14), ΔU, including p control inputs, need to be optimized, while only the first one is used for control. This will occupy a great deal of computation resource and result in very low computational efficiency, especially with respect to the fast applications.

In order to reduce the computational burden of Eq. (14), we propose here a 'step plan' technique,

$$\Delta u_{t+i+1} = \beta \Delta u_{t+i}$$

(17)

where β is an $m \times m$ diagonal matrix presenting the length of one step, which will be a parameter to be selected. Then, we can simplify Eq. (14) by only calculating the unknown control, which has smaller dimensions.

where $Im \times m$ is an $m \times m$ unit matrix. Thus, the number of the unknown control input vector (from current time t to the future time $t+p-1$) is reduced from p to 1, and the dimension of predictive matrix is changed from $pl \times pm$ to $pl \times m$. This reduction brings low computer memory consuming and simplifies the receding horizon optimization in the following calculation.

To complete the horizon optimization and obtain the control input, the cost function of the stationary increment predictive control is designed as:

$$J = (R_t - \hat{Y}_t)^T W (R_t - \hat{Y}_t) + \Delta u_t^{\ T} \lambda \Delta u_t$$

(19)

Where $R_t = \left(r_{t+k+1}^T \quad r_{t+k+2}^T \quad \cdots \quad r_{t+k+p}^T \right)^T$, $W \in R^{lp \times lp}$ is the weight matrix for tracking error, and $\lambda \in Rm \times m\lambda \in Rm \times m$ is the weight matrix of the control increment.

In order to minimize the cost function of Eq. (19), we can calculate the control vector as follows:

$$\Delta u_t = (G_2^T W G_2 + \lambda)^{-1} G_2^T W (R_t - Y_t^1)$$
$$= K_f (R_t - Y_t^1)$$

where $K_f = (G_2^T W G_2 + \lambda)^{-1} G_2^T W$ can be completed offline.

Consequently, the proposed stationary increment predictive controller (SIPC) can be designed as followings.

Step I: Make increment prediction

Based on the current and history measure value, use Eqs. (13-15) to obtain the prediction for future output $\hat{Y}_t$ and initial plan point

Step II: Plan for the set-point input

Use Eq. (12) to plan the future set-points, and obtain

$$R_t = \left(r_{t+k+1}^T \quad r_{t+k+2}^T \quad \cdots \quad r_{t+k+p}^T \right)^T$$

Step III: Receding horizon optimization

Calculate the control increment Δu_t, based on Eq. (20).

Step IV: Control implementation

Current control input $u_t = u_{t-1} + \Delta u_t$, which is used as the control to the real plant. After that, go back to step I at the next time instant.

Thus, for real implementation, only the prediction of Eq. (13-15), the intenerating of Eq. (12), and the control law (20) need to be calculated online, thus the real time computation load, and steady tracking error are both reduced greatly compared with GPC, and the real test in section V has shown its feasibility.

The model error, problem 3), will be compensated by an online optimal strategy, which will be described later.

4.3. Optimal strategy for model error compensation

In order to compensate the model error in Eq. (1), the control vector has to match the following equation, which can be directly obtained from Eq. (1):

$$B_d U_t + B_f f_t = B_d U_t^0$$

$$(21)$$

where U^0_t is the control vector need to be calculated by the predictive controller in section 4.2, designed based on the original model (1) without the model error f.

The control input at sampling time t cannot be solved directly from Eq. (21), because:

1. Eq. (21) is difficult to be implemented because the dimension of U_t is less than that of f_t. Thus, only the approximate solution can be obtained with respect to (21);
2. f_t is actually an uncertainty set, an static optimal problem must be considered. Thus, we introduce the following cost function with quadratic form to solve the above problem 1).

$$U_t^* = \arg \min_{U_t} J_t(U_t)$$

$$(22)$$

$$J_t(U_t) \triangleq \left(B_d U_t + B_f f_t - B_d U_t^0 \right)^T H \left(B_d U_t + B_f f_t - B_d U_t^0 \right)$$

where H is a weight matrix, which can be selected.

On the other hand, f_t is obtained from the ASMF algorithm introduced in section III, thus its convergence is very important for the validity of the whole controller. Actually, the convergence of ASMF algorithm is also influenced by the control action U_t. This is because the stability of the ASMF can be represented by the filter parameter δ_t, while δ_t in Eq. (3) can be rewritten as follows,

$$\delta_t = 1 - (Y_t - C_d^a \hat{X}_{t|t-1}^a)^T W_t^{-1} (Y_t - C_d^a \hat{X}_{t|t-1}^a)$$
$$= 1 - (Y_{t+1} - C_d^a (A_d^a \hat{X}_{t|t}^a + B_d^a U_t))^T$$
$$W_t^{-1} (Y_{t+1} - C_d^a (A_d^a \hat{X}_{t|t}^a + B_d^a U_t))]$$

(23)

In [19], it has been shown the stability of the ASMF can be represented by the filter parameter δ_t, i.e., the ASMF is stable when $\delta_t > 0$.

Firstly, define

$$J_t^\delta (U_t, Y_{t+1}) \triangleq (Y_{t+1} - C_d^a (A_d^a \hat{X}_{t|t}^a + B_d^a U_t))^T$$
$$W_t^{-1} (Y_{t+1} - C_d^a (A_d^a \hat{X}_{t|t}^a + B_d^a U_t))]$$

(24)

Thus, from Eq. (23), in order to maintain $\delta_{t+1} > 0$, the maximum value of $J_t^\delta (U_t, Y_{t+1})$ with respect to $\hat{X}_{t|t}^a$ should be less than or equal to 1, i.e.,

$$J_t^{\delta^*} (U_t, Y_{t+1}) = \max_{\hat{X}_{t|t}^a} J_t^\delta (U_t, Y_{t+1})$$
$$= \max_{\hat{X}_{t|t}^a} \{ [Y_{t+1} - C_d^a (A_d^a \hat{X}_{t|t}^a + B_d^a U_t)]^T W_t^{-1} [Y_{t+1} - C_d^a (A_d^a \hat{X}_{t|t}^a + B_d^a U_t)] \} \le 1$$

(25)

In general, larger δ_t often means more rapid convergence of ASMF algorithm. That is, we should select an U_t to make $J_t^{\delta^*}(U_t,Y_{t+1})$ small as far as possible, that is,

$$J_t^{\delta^*} (Y_{t+1}) = \min_{U_t} J_t^{\delta^*} (U_t, Y_{t+1})$$

(26)

We introduce the following cost function $J_t(U_t)$ with consideration of both (22) and (25) at the same time:

$$^*U_t = \arg \min_{U_t} \bar{J}_t(U_t)$$

$$\bar{J}_t(U_t) \triangleq J_t(U_t) + \alpha J_t^{\delta^*}(U_t, Y_{t+1}) \tag{27}$$

where $\alpha = 1 - \delta_t \in R$ are the positive definite weight matrix. To minimize $J_t(U_t)$, considering , the control can be obtained at $\partial J_t(U_t)/\partial U_t = 0$, i.e.,

$$\partial J_t(U_t)/\partial U_t = 2(MU_t + N) \tag{28}$$

where

$$M = B_d^T H B_d + \alpha B_d^{aT} C_d^{aT} W_t^{-1} C_d^a B_d^a$$

$$N = B_d^T H (B_f f_t - B_d U_t^0) - \alpha B_d^{aT} C_d^{aT} W_t^{-1} Y_{t+1}$$

Here H can be selected as $H = \delta_t C_d^T C_d$. Thus, we can obtain the optimal control that minimizes $J_t(U_t)$ as:

$$U_t(Y_{t+1}) = -M^{-1}N$$
$$= (B_d^T H B_d + \alpha B_d^{aT} C_d^{aT} W_t^{-1} C_d^a B_d^a)^{-1}$$
$$\left[\alpha B_d^{aT} C_d^{aT} W_t^{-1} Y_{t+1} - B_d^T H (B_f f_t - B_d U_t^0) \right] \tag{29}$$

For the unknown measurement at time t+1 in Eq. (24), we consider that the control system is stable, so, $Y_{t+1} \in \Delta(Y_t)$. Here, $\Delta(Y_t)$ is the elliptical domain of Y_t. Because $J_t^\delta(U_t, Y_{t+1})$ in Eq. (25) is positive definite, its maximum value point must be on the boundary, which can be estimated by the

ASMF. Thus, we first define array S_t^i to include the estimate of the i-th element's two boundary endpoints as

$$S_t^i \triangleq \left\{ \hat{Y}_{t+1}^i \mid \{Y_t^i + (-1)^{h}(\max_{j_1=\{\pm\sqrt{p_{11}},l=1,\dots,13\}} \left|C_d Col\{j\}\right|_i)\} \right\}$$

(30)

where Y_t^i is the i-th element in the vector Y_t, $\hat{Y}_{t+1}^i$, is the corresponding output Y_{t+1}'s endpoints estimation. For set S_t^i, $i \in \{1,2,\dots,8\}$, and h is 0 or 1 for every i, $\left|\bullet\right|_i$ is the operator for absolute value of the i-th element in vector $\bullet$, and the function $Col\{j\}$ is defined as follows:

$$Col\{j\} = \begin{pmatrix} j_1 & \dots & j_{13} \end{pmatrix}^T$$

(31)

Then, we define a set S_t to describe all possible endpoint vector of the Y_{t+1} as

$$S_t \triangleq \left\{ \hat{Y}_{t+1}^{EP} \mid \begin{pmatrix} S_t^1 & \dots & S_t^{13} \end{pmatrix} \right\}$$

(32)

where $\hat{Y}_{t+1}^{EP}$ is the possible endpoint (EP) for output Y_{t+1} at next sampling time $t+1$.

Thus, the proposed active modeling based predictive controller can be implemented by using the following steps:

Step I: Make increment prediction

Based on the current estimated state $\hat{X}_{t|t}^a$, use the stationary increment predictive controller, as in section 4.2, to obtain the nominal control input U_t^0;

Step II: Model error estimation and elimination

Based on U_t^0, compute the optimal control input $^*U_t$:

Estimate the values and boundaries of state X_t and model error f_t, using ASMF in (3);

Calculate the corresponding $U_t(\hat{Y}_{t+1}^{EP})$ for every $\hat{Y}_{t+1}^{EP}$ in set S_t by Eq. (29);

For every $U_t(\hat{Y}_{t+1}^{EP})$ in step 1), use Eq. (24) to obtain the maximum of function

$$J_t^\delta(U_t(\hat{Y}_{t+1}^{EP}), \hat{Y}_{t+1}^{EP}), \text{ and get the } {}^*\hat{Y}_{t+1}^{EP} \text{ to let}$$

$${}^*\hat{Y}_{t+1}^{EP} = \arg \underset{\hat{Y}_{t+1}^{EP} \in S_t}{\operatorname{Max}} \left\{ J_t^\delta \{ U_t(\hat{Y}_{t+1}^{EP}), \hat{Y}_{t+1}^{EP} \} \right\};$$

The corresponding $U_t({}^*\hat{Y}_{t+1}^{EP})$ is the optimal control ${}^*U_t$ at time t, i.e. ${}^*U_t = U_t({}^*\hat{Y}_{t+1}^{EP})$.

Step III: Receding horizon strategy

Go back to step I at the next time instant.

5. FLIGHT TEST

5.1. Flight test platform

All flight tests are conducted on the Servoheli-40 setup, which was developed in the State Key Laboratory, SIACAS. It is equipped with a 3-axis gyro, a 3-axis accelerometer, a compass and a GPS. The sensory data can be sampled and stored into an SD card through an onboard DSP. Tab.1 shows the physical characteristics of SERVOHELI-40 small-size helicopter. More details of this experimental platform can be found in [24].

Figure 2. SERVOHELI-40 small-size helicopter platform

Table 1. Physical characteristics of SERVOHELI-40 small-size helicopter

Length	2.12m
Height	0.73m
Main rotor diameter	2.15m
Stabilizer bar diameter	0.75m
Rotor speed	1450rpm
Dry weight	20kg
Engine	2-stroke, air cooled
Flight time	45 min

5.2. Experiment For The Verification Of Model Error Estimate When Mode-Change

We use the identified hovering parameters, through frequency estimate [25], as the nominal model for hovering dynamics of the ServoHeli-40 platform. The model accuracy is verified in hovering mode (speed less than 3m/s) and cruising mode (speed more than 5m/s), the results for lateral velocity are shown in Fig.3a.

Fig. 3 further shows the model difference due to mode change, where the red lines are the results calculated by the identified model with the inputs of hovering and cruising actuations, respectively, and blue lines are the measurements of the onboard sensors. Comparison shows that the hovering model outputs match the hovering state closely, but clear differences occur while being compared to the cruising state, even though the cruising actuations are used as the model inputs. This is the model error when flight mode is changed.

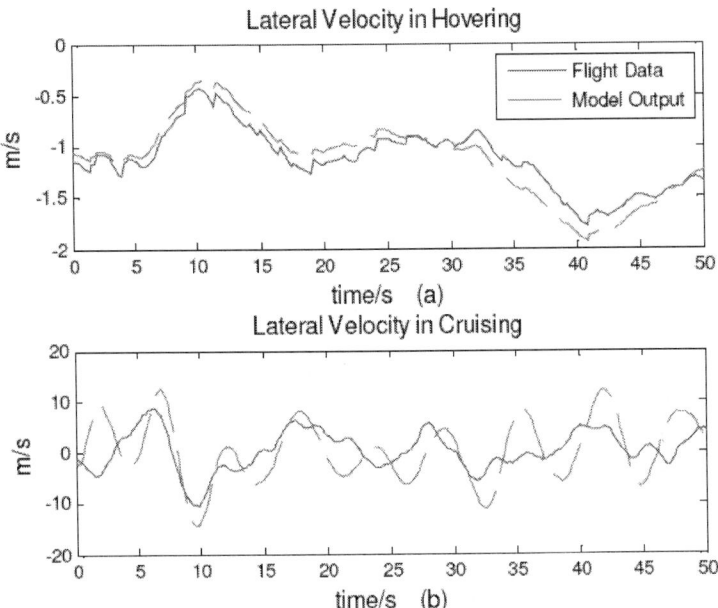

Figure 3. Model difference due to mode change: (a) hovering conditions; (b) cruising conditions

To verify the accuracy of the estimate of the model error, described in Fig.3, the following experiment is designed:

1. Actuate the longitudinal control loop to keep the speed more than 5 meter per second;
2. Get the lateral model error value and boundaries through ASMF, and add them to the hovering model we built above;
3. Compare the model output before and after compensation for model error.

This process of experiment can be described by Fig.4, and the results are shown in Fig.5. Fig.5a shows that model output (red line) cannot describe the cruising dynamics due to the model error when 'mode-change', similar with Fig.3b; however, after compensation, shown in Fig.5b, the model output (red line) is very close with real cruising dynamics (blue line), and the uncertain boundaries can include the changing lateral speed, which mean that the proposed estimation method can obtain the model error and range accurately by ASMF when mode-change.

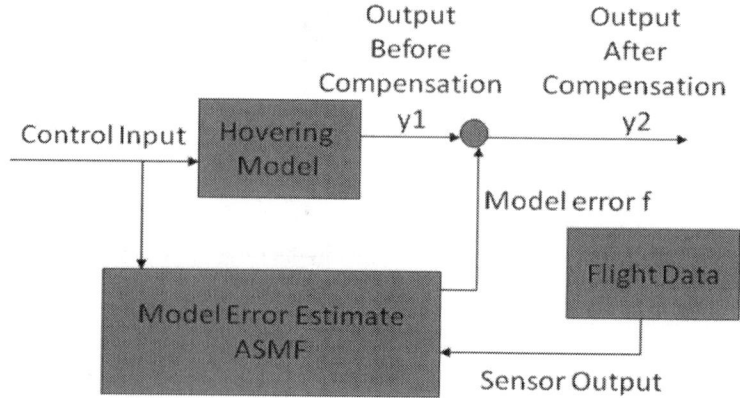

Figure 4. The experiment process for model-error estimate

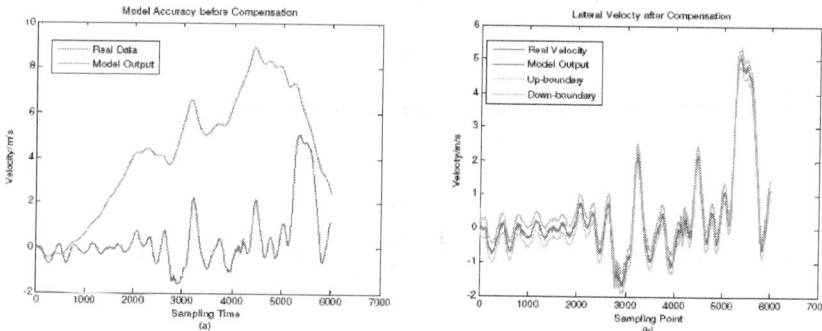

Figure 5. Model output before/after compensation: (a) before compensation; (b) after compensation

5.3. Flight Experiment For The Comparison of Gpc Sipc and Amsipc When Sudden Mode-Change

In Section 5.2, the model-error occurrence and the accuracy of the proposed method for estimation are verified. So, the next is the performance of the proposed controller in real flight. In this section, the performance of the modified GPC (Generalized Predictive Control, designed in Section 4.1), SIPC (Stationary Increment Predictive Control, designed in Section 4.2) and AMSIPC (Active Modeling Based Stationary Increment Predictive Control, designed in Section 4.3), are tested in sudden mode-change, and are compared with each other on the ServoHeli-40 test-bed. To complete this mission, the following experimental process is designed:

1. Using large and step-like reference velocity, red line in Fig.6-8, input it to longitudinal loop, lateral loop and vertical loop;

2. Based on the same inputted reference velocity, using the 3 types of control method, GPC, SIPC and AMSIPC to actuate the helicopter to change flight mode quickly;
3. Record the data of position, velocity and reference speed for the 3 control loops, and obtain reference position by integrating the reference speed;
4. Compare errors of velocity and position tracking of GPC, SIPC and AMSIPC, executively, in this sudden mode-change flight.

GPC, SIPC and AMSIPC are all tested in the same flight conditions, and the comparison results are shown in Figs. 6-8. We use the identified parameters in Section 5.2 to build the nominal model, based on the model structure in Appendix A, and parameters' selection in Appendix C for controllers

It can be seen that, when the helicopter increases its longitudinal velocity and changes flight mode from hovering to cruising, GPC (brown line) has a steady velocity error and increasing position error because of the model errors. SIPC (blue line) has a smaller velocity error because it uses increment model to reject the influence of the changing operation point and dynamics' slow change during the flight. The prediction is unbiased and obtains better tracking performance, which is verified by Theorem. However, the increment model may enlarge the model errors due to the uncertain parameters and sensor/process noises, resulting in the oscillations in the constant velocity period (clearly seen inFig.6&7) because the error of its prediction is only unbiased, but not minimum variance. While for AMSIPC (green line), because the model error, which makes the predictive process non-minimum variance, has been online estimated by the ASMF and compensated by the strategy in section 4.3, the proposed AMSIPC successfully reduces velocity oscillations and tracking errors together.

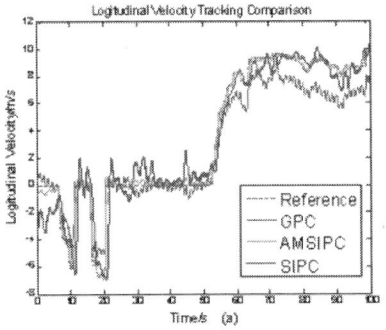
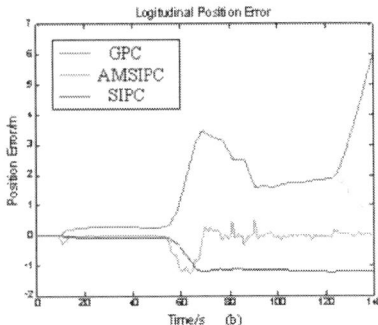

Figure 6. Longitudinal tracking results: (a) velocity; (b) position error (<50s hovering, >50s cruising)

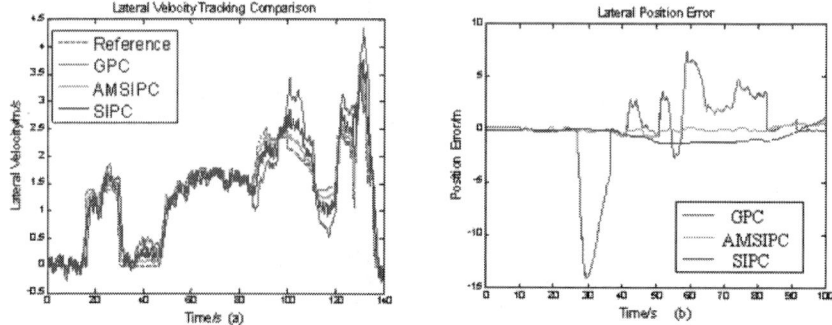

Figure 7. Lateral tracking results: (a) velocity; (b) position error (25s~80s cruising, others hovering)

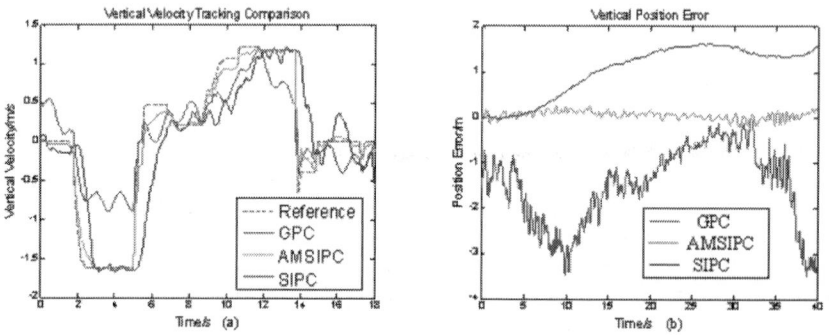

Figure 8. Vertical tracking results: (a) velocity; (b) position error (<5s hovering; >5s cruising)

6. CONCLUSION

An active model based predictive control scheme was proposed in this paper to compensate model error due to flight mode change and model uncertainties, and realize full flight envelope control without multi-mode models and mode-dependent controls.

The ASMF was adopted as an active modeling technique to online estimate the error between reference model and real dynamics. Experimental results have demonstrated that the ASMF successfully estimated the model error even though it is both helicopter dynamics and flight-state dependent.In order to overcome the aerodynamics time-delay, also with the active estimation for optimal compensation, an active modeling based stationary increment predictive controller was designed and analyzed.

The proposed control scheme was implemented on our developed ServoHeli-40 unmanned helicopter. Experimental results have demonstrated clear improvements over the normal GPC without active modeling enhancement when sudden mode-change happens.

It should be noted that, at present, we have only tested the control scheme with respect to the flight mode change from hovering to cruising, and vice versa. Further mode change conditions will be flight-tested in near future.

APPENDIX

A. Helicopter dynamics

A helicopter in flight is free to simultaneously rotate and translate in six degrees of freedom. Fig. 9shows the helicopter variables in a body-fixed frame with origin at the vehicle's center of gravity.

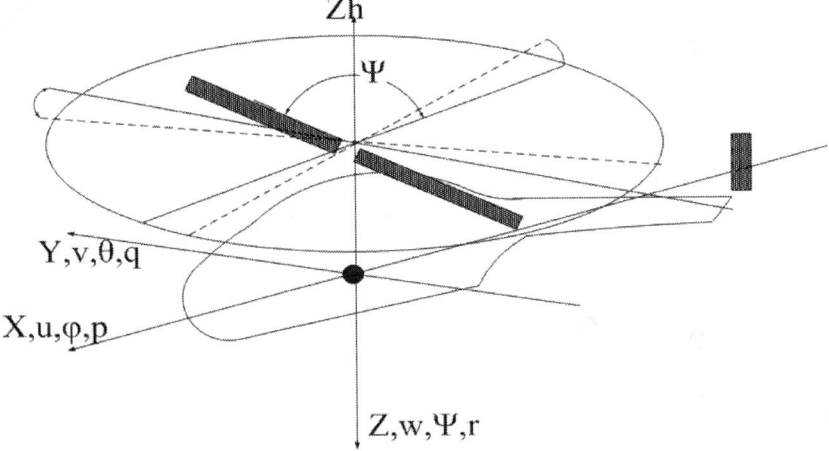

Figure 9. Helicopter with its body-fixed reference frame

Ref. [18] developed a semi-decoupled model for small-size helicopter, i.e.,

$$
\begin{pmatrix} \delta\dot{u} \\ \delta\dot{q} \\ \delta\dot{\theta} \\ \dot{a} \\ \dot{c} \end{pmatrix} = \begin{pmatrix} X_u & 0 & -g & X_a & 0 \\ M_u & 0 & 0 & M_a & 0 \\ 0 & 1 & 0 & 0 & 0 \\ 0 & -1 & 0 & -1/\tau_f & A_c/\tau_f \\ 0 & -1 & 0 & 0 & -1/\tau_f \end{pmatrix} \begin{pmatrix} \delta u \\ \delta q \\ \delta\theta \\ a \\ c \end{pmatrix} + \begin{pmatrix} X_{lon} & X_{lat} \\ M_{lon} & M_{lat} \\ 0 & 0 \\ A_{lon} & A_{lat} \\ C_{lon} & C_{lat} \end{pmatrix} \begin{pmatrix} \delta_{lon} \\ \delta_{lat} \end{pmatrix}, \text{ i.e.,}
$$

$$
\begin{cases} \dot{X}_{lon} = A_{lon}\delta X_{lon} + B_{lon}\delta u_{lon} \\ y_{lon} = \begin{pmatrix} I_{3\times3} & 0_{3\times2} \end{pmatrix} \delta X_{lon} = C_{lon}\delta X_{lon} \end{cases}
$$

$$(A-1)$$

$$\begin{pmatrix} \delta \dot{v} \\ \delta \dot{p} \\ \delta \dot{\varphi} \\ \dot{b} \\ \dot{d} \end{pmatrix} = \delta \dot{X}_{lat} = \begin{pmatrix} Y_u & 0 & g & Y_a & 0 \\ L_u & 0 & 0 & L_a & 0 \\ 0 & 1 & 0 & 0 & 0 \\ 0 & -1 & 0 & -1/\tau_f & B_d/\tau_f \\ 0 & -1 & 0 & 0 & -1/\tau_f \end{pmatrix} \begin{pmatrix} \delta v \\ \delta p \\ \delta \varphi \\ b \\ d \end{pmatrix} + \begin{pmatrix} Y_{lon} & Y_{lat} \\ L_{lon} & L_{lat} \\ 0 & 0 \\ B_{lon} & B_{lat} \\ D_{lon} & D_{lat} \end{pmatrix} \begin{pmatrix} \delta_{lon} \\ \delta_{lat} \end{pmatrix}, \text{ i.e.,}$$

$$\begin{cases} \dot{X}_{lat} = A_{lat}\delta X_{lat} + B_{lat}\delta u_{lat} \\ y_{lat} = \begin{pmatrix} I_{3\times3} & 0_{3\times2} \end{pmatrix}\delta X_{lat} = C_{lat}\delta X_{lat} \end{cases}$$

(A-2)

$$\begin{pmatrix} \delta \dot{w} \\ \delta \dot{r} \\ \delta \dot{r}_{fb} \end{pmatrix} = \delta \dot{X}_{yaw-heave} = \begin{pmatrix} Z_w & Z_r & 0 \\ N_w & N_r & -N_{ped} \\ 0 & K_r & -K_{rfb} \end{pmatrix} \begin{pmatrix} \delta w \\ \delta r \\ \delta r_{fb} \end{pmatrix} + \begin{pmatrix} Z_{ped} & Z_{col} \\ N_{ped} & N_{col} \\ 0 & 0 \end{pmatrix} \begin{pmatrix} \delta_{ped} \\ \delta_{col} \end{pmatrix}, \text{ i.e.}$$

$$\begin{cases} \dot{X}_{yaw-heave} = A_{yaw-heave}\delta X_{yaw-heave} + B_{yaw-heave}\delta u_{yaw-heave} \\ y_{yaw-heave} = \begin{pmatrix} I_{2\times2} & 0_{2\times1} \end{pmatrix}\delta X_{yaw-heave} = C_{yaw-heave}\delta X_{yaw-heave} \end{cases}$$

(A-3)

where δu, δv, δw are longitudinal, lateral and vertical velocity, δp, δq, δr are roll, pitch and yaw angle rates, $\delta \varphi$ and $\delta \theta$ are the angles of roll and pitch, respectively, a and b are the first harmonic flapping angle of main rotor, c and d are the first harmonic flapping angle of stabilizer bar, δr_{fb} is the feedback control value of the angular rate gyro, δ_{lat} is the lateral control input, δ_{lon} is the longitudinal control input, δ_{ped} is the yawing control input, and δ_{col} is the vertical control input. All the symbols except gravity acceleration g in A_{lon}, A_{lat}, $A_{yaw-heave}$, B_{lon}, B_{lat} and $B_{yaw-heave}$ are unknown parameters to be identified. Thus, all of the states and control inputs in (33), (34) and (35) are physically meaningful and defined in body-axis.

B. Proof for the predictive theorem

Proof:
Assume the real dynamics is described as:

$$X_{t+1} = A_{dr}X_t + B_{dr}U_{t-k} + W_t$$

(B-1)

which is different from the reference model of Eq. (11). In Eq. (B-1), X_t is system state, A_{dr} is the system matrix, B_{dr} is the control matrix, U_t is control input, W_t is process noise. The one-step prediction, according to Eq. (B-1), can be obtained by Eq. (13-14),

$$\hat{X}_{t|t+1} = X_t + A_d \Delta X_t + B_d \Delta U_{t-k}$$
$$= A_{dr} X_{t-1} + B_{dr} U_{t-1-k} + W_{t-1}$$
$$+ A_d \Delta X_t + B_d \Delta U_{t-k} \tag{B-2}$$

And

$$E\left\{ X_{t+1} - \hat{X}_{t+1|t} \right\}$$
$$= E\{ A_{dr} X_t + B_{dr} U_{t-k} + W_t$$
$$- (A_{dr} X_{t-1} + B_{dr} U_{t-1-k} + W_{t-1} + A_d \Delta X_t + B_d \Delta U_{t-k}) \}$$
$$= E\{ (A_{dr} - A_d) \Delta X_t + (B_{dr} - B_d) \Delta U_{t-k} + \Delta W_t \} \tag{B-3}$$

According to condition 1) and 2), prediction is bounded, then,

$$\left\| X_{t+1} - \hat{X}_{t+1|t} \right\| < +\infty$$

and, when the system of Eq. (36) works around a working point in steady state, the mean value of control inputs and states should be constant, so we can obtain:

$$E\left\{ X_{t+1} - \hat{X}_{t+1|t} \right\}$$
$$= (A_{dr} - A_d) E\{\Delta X_t\} + (B_{dr} - B_d) E\{\Delta U_{t-k}\} + E\{\Delta W_t\}$$
$$= (A_{dr} - A_d) \bullet 0 + (B_{dr} - B_d) \bullet 0 + 0 = 0 \tag{B-4}$$

Eq. (B-4) indicates that the one step prediction of Eq. (B-2) is unbiased. Assuming that prediction at time i-1 is unbiased, i.e.,

$$E\{ X_{t+i-1} - \hat{X}_{t+i-1|t} \} = 0 \tag{B-5}$$

for the prediction at time i, there is

$$
\begin{aligned}
E&\{X_{t+i} - \hat{X}_{t+i|t}\} \\
&= E\{A_{dr}X_{t+i-1} + B_{dr}U_{t+i-1-k} + W_{t+i-1} \\
&\quad - (\hat{X}_{t+i-1|t} + A_d\hat{\Delta X}_{t+i-1|t} + B_d\Delta U_{t+i-1-k})\} \\
&= E\{A_{dr}X_{t+i-1} + B_{dr}U_{t+i-1-k} + W_{t+i-1} - X_{t+i-1} \\
&\quad + (X_{t+i-1} - \hat{X}_{t+i-1|t}) - W_{t+i-2} \\
&\quad - A_d\hat{\Delta X}_{t+i-1|t} - B_d\Delta U_{t+i-1-k}\} \\
&= E\{A_{dr}\Delta X_{t+i-1} - A_d\hat{\Delta X}_{t+i-1|t} + \\
&\quad (B_{dr} - B_d)\Delta U_{t+i-1-k} + \Delta W_{t+i-1}\} \\
&= (A_{dr} - A_d)E\{\Delta X_{t+i-1}\} + \\
&\quad (B_{dr} - B_d)E\{\Delta U_{t+i-1-k}\} + E\{\Delta W_{t+i-1}\} \\
&= (A_{dr} - A_d)\bullet 0 + (B_{dr} - B_d)\bullet 0 + 0 = 0
\end{aligned}
\tag{B-6}
$$

Therefore, the prediction at time i is also unbiased.

C. Parameters' selection for estimate and control in flight experiment

1.　　　For Modeling

The identification results for hovering dynamics are listed in Tab.D-1.

2.　　　For ASMF

$$
Q = \begin{pmatrix} 0.01I_{13\times13} & 0_{13\times13} \\ 0_{13\times13} & 0.1I_{13\times13} \end{pmatrix}, \quad R = 0.01I_{8\times8}
$$

where $I_{m\times m}$ is the m×m unit matrix and $0_{m\times n}$ is the m×n zero matrix.

Table D-1. The parameters of hovering model

Longitudinal Loop		Lateral	Loop	Vertical	Loop
Para.	Val.	Para.	Val.	Para.	Val.
Xu	0.2446	Yv	-0.0577	Zw	1.666
Xa	-4.962	Yb	9.812	Zr	-3.784
Xlat	-0.0686	Ylat	-1.823	Zped	2.304
Xlon	0.0896	Ylon	2.191	Zcol	-11.11
Mu	-1.258	Lv	15.84	Yaw Loop	
Ma	46.06	Lb	126.6	Para.	Val.
Mlat	-0.6269	Llat	-4.875	Nw	-0.027
Mlon	3.394	Llon	28.64	Nr	-1.087
Ac	0.1628	Bd	-1.654	Nrfb	-1.845
Alat	-0.0178	Blat	0.04732	Nped	1.845
Alon	-0.2585	Blon	-9.288	Ncol	-0.972
Clat	2.238	Dlat	-0.7798	Kr	-0.040
Clon	-4.144	Dlon	-5.726	Krfb	-2.174
tf	0.5026	ts	0.5054		

3.   For GPC

$$p = 10 \, , \; \gamma = 2.32I_{40\times40} \, , \; k = 10$$

4.   For SIPC

$$p = 10 \, , \; \gamma = 2.32I_{4\times4} \, , \; \alpha = 0.99I_{8\times8}$$

$$W = I_{80\times80} \, , \; k = 10 \, , \; \beta = 0.8I_{4\times4}$$

5.   For AMSIPC

$$p = 10 \, , \; \gamma = 2.32I_{4\times4} \, , \; \alpha = 0.99I_{8\times8}$$

$$W = I_{80\times80} \, , \; k = 10 \, , \; \beta = 0.8I_{4\times4} \, , \; H = I_{13\times13}$$

REFERENCES

1. Tischler M.B., "Frequency-domain Identification 15 XV-15 Tilt-rotor Aircraft Dynamics in Hovering Flight," Journal of the American Helicopter Society, 30 2), 384 , 1985.

2. M. B. Tischler, M. G. Cauffman, ". Frequency-Response, for. Method, System. Rotorcraft, Flight. Identification, to. B. O. Application, 05 Coupled Rotor/Fuselage Dynamics," Journal of the American Helicopter Society, 37 3), 317, 1992.

3. J. W. Fletcher, ". Identification, 60 UH-60 Stability Derivative Models in Hover from Flight Test Data," Journal of the American Helicopter Society, 40 1), 820 , 1995.

4. B. Mettler, M. B. Tischler, T. Kanade, ". System, of. Identification, Unmanned. Small-Size, Dynamics,". Helicopter, Helicopter. American, Society, 55 Annual Forum Proceedings, 2 17061717, Montreal, Quebec, Canada, May 25-27, 1999.

5. V. Gavrilets, B. Metlter, E. Feron, ". Nonlinear, for. a. Model, Acrobatic. Small-scale, Proceedings. Helicopter,", the. of, Institute. American, Aeronautics. of, Navigation. Guidance, Conference. Control, 8 8 Montreal, Quebec, Canada, August 6-9, 2001.

6. M. Massimiliano, S. Valerio, ". A. Full, Small. Envelope, Aircraft. Commercial, Control. Flight, Using. Design, Proportional. Multivariable-Integral, I. E. E. E. Control,", on. Transactions, Systems. Control, Vol. Technology, 161 (1), 169176, January, 2008.

7. M. Voorsluijs, A. Mulder, . Parameter-dependent, control. robust, a. for, U. A. V. rotorcraft, A. I. A. A. Guidance, Navigation, Conference. Control, Exhibit, 111 111 San Francisco, California, USA, August 15-18, 2005.

8. Bijnens B., Chu Q.P. and Voorsluijs M., "Adaptive feedback linearization flight control for a helicopter UAV," AIAA Guidance, Navigation, and Control Conference and Exhibit, 110 110 San Francisco, California, USA, August 15-18, 2005.

9. Kahveci N.E., Ioannou P.A., Mirmirani M.D., "Adaptive LQ Control With Anti-Windup Augmentation to Optimize UAV Performance in Autonomous Soaring Applications," IEEE Transactions on Control Systems Technology, Vol. 164 691707 , 2008

10. MacKunis W., Wilcox Z.D., Kaiser M.K., Dixon W.E., "Global Adaptive Output Feedback Tracking Control of an Unmanned Aerial Vehicle," IEEE Transactions on Control Systems Technology, Vol. 186 13901397, 2010.

11. M. L. Cummings, P. J. Mitchell, ". Predicting, Capacity. Controller, Supervisory. in, of. Control, U. A. Multiple, I. E. E. E. Systems,", on. Transactions, Man, Part. A. Cybernetics, Systems, Vol. Humans, 382 451460, 2008.

12. Jiang X., Han Q.L., "On guaranteed cost fuzzy control for nonlinear systems with interval time-varying delay," Control Theory & Applications, IET, Vol. 16 17001710, 2007.

13. Natori K., Oboe R., Ohnishi, K., "Stability Analysis and Practical Design Procedure of Time Delayed Control Systems With Communication Disturbance Observer," IEEE Transactions on Industrial Informatics, Vol. 43 185197 , 2008.

14. Haykin, and De Freitas N., "Special Issue on Sequential State Estimation," Proceedings of the IEEE, 92 923 423574 , 2004.

15. D. Lerro, Y. K. Bar-Shalom, ". Tracking, Debiased. with, Converted. Consistent, E. K. Measurements, EKF," IEEE Transactions on Aerosp. Electron.System, 29 10151022, 1993.

16. S. Julier, J. Uhlmann, ". Unscented, filtering, estimation,". nonlinear, of. Proceedings, I. E. E. E. the, Vol, 923 401422, 2004.

17. Song Q., Jiang Z., and Han J. D., "UKF-Based Active Model and Adaptive Inverse Dynamics Control for Mobile Robot," IEEE International Conference on Robotics and Automation, 2007.

18. J. S. Shamma, K. Y. Tu, ". Approximate, observers. set-valued, nonlinear. for, I. E. E. E. systems,", on. Transactions, Control. Automatic, Vol, 425 648658, 1997.

19. Zhou B., Han J.D. and Liu G., "A UD factorization-based nonlinear adaptive set-membership filter for ellipsoidal estimation," International Journal of Robust and Nonlinear Control, 18 16), 15131531, November 10, 2007.

20. Scholte E., Campbell M.E., "Robust Nonlinear Model Predictive Control With Partial State Information," Control Systems Technology, IEEE Transactions on, Vol. 164 636651, 2008.

21. B. C. Ding, Y. G. Xi, ". A. Synthesis, of. Approach, Constrained. On-line, Model. Robust, Control.". Predictive, Automatica, 40 401 163167, 2004.

22. J. L. Crassidis, ". Robust, of. Control, Systems. Nonlinear, Model. Using-Error, Synthesis,". Control, of. Journal, control. guidance, Vol. dynamics, 224 (4), 595601 , 1999.

23. K. Gregor, S. Igor, ". Tracking-error, Predictive. Model-based, for. Control, Robots. Mobile, real. in, Robotics. time.", Systems. Autonomous, 55 55 7 460469 , 2007.

24. Qi J.T., Song D.L., Dai. L., Han J.D., "The ServoHeli-20 Rotorcraft UAV Project," International Conference on Mechatronics and Machine Vision in Practice, Auckland, New Zealand, pp.92-96, 2008.

25. Song D.L., Qi J.T., Dai. L., Han J.D. and Liu G., "Modeling a Small-size Unmanned Helicopter Using Optimal Estimation in The Frequency Domain," International Conference on Mechatronics and Machine Vision in Practice, Auckland, New Zealand, December 24 97102, 2008.

26. Song D.L., Qi J.T. and Han J.D., "Model Identification and Active Modeling Control for Small-Size Unmanned Helicopters: Theory and Experiment," AIAA Guidance Navigation and Control, Toronto, Canada, AIAA-2010-7858, 2010.

CHAPTER 5

Intelligent and Predictive Vehicular Networks

Schmidt Shilukobo Chintu, Richard Anthony, Maryam Roshanaei, Constantinos Ierotheou

Faculty of Architecture, Computing and Humanities, School of Computing & Mathematical Sciences, University of Greenwich, London, UK

ABSTRACT

Seeking shortest travel times through smart algorithms may not only optimize the travel times but also reduce carbon emissions, such as CO_2, CO and Hydro-Carbons. It can also result in reduced driver frustrations and can increase passenger expectations of consistent travel times, which in turn points to benefits in overall planning of day schedules. Fuel consumption savings are another benefit from the same. However, attempts to elect the shortest path as an assumption of quick travel times, often work counter to the very objective intended and come with the risk of creating a "Braess Paradox" which is about congestion resulting when several drivers attempt to elect the same shortest route. The situation that arises has been referred to as the price of anarchy! We propose algorithms that find multiple shortest paths between an origin and a destination. It must be appreciated that these will not yield the exact number of Kilometers travelled, but favourable weights in terms of travel times so that a reasonable allowable time difference between the multiple shortest paths is attained when the same Origin and Destinations are considered and favourable responsive routes are determined as variables of traffic levels and time of day. These routes are selected on the paradigm of route balancing, re-routing algorithms and traffic light intelligence all coming together to result in optimized consistent travel times whose benefits are evenly spread to all motorist, unlike the Entropy balanced k shortest paths (EBkSP) method which favours some motorists on the basis of urgency. This paper proposes a Fully Balanced MultipleCandidate shortest path (FBMkP) by which we model in SUMO to overcome the computational overhead of assigning priority differently to each travelling vehicle using intelligence at intersections and other points on the vehicular network. The FBMkP opens up traffic by fully balancing the whole network so as to benefit every motorist. Whereas the EBkSP reserves some routes for cars on high priority, our algorithm distributes the benefits of smart routing to all vehicles on the network and serves the road side units such as induction loops and detectors from having to remember the urgency of each vehicle. Instead, detectors and induction loops simply have to poll the destination of the vehicle and not any urgency factor. The minimal data being processed significantly reduce computational times and the benefits all vehicles. The multiple-candidate shortest paths selected on the basis of current traffic status on each possible route increase the efficiency. Routes are fewer than vehicles so possessing weights of routes is smarter than processing individual vehicle weights. This is a multi-objective function project where improving one factor such as travel times improves many more cost, social and environmental factors.

Keywords: Simulation of Urban Mobility SUMO, Duarouter, Fully Balanced Multiple-Candidate Shortest Paths (FBMKP), E1 Induction Loop, E3 Detector, Re-Routing, Braess Paradox, Traffic Control Intelligent (TraCI), Partially Re-Routed, Shortest Path Method, Traffic Light Control FBMKP

1. INTRODUCTION

The problem statement has its focus on achieving an intelligent, predictive and responsive traffic model by improving the response time between road side infrastructure and routed vehicles on a least cost approach. The approach of balancing the network without prioritizing any set of vehicles frees the responsibility of roadside units to any specific vehicles, and instead, the units along with traffic lights in another version of the fully balanced algorithm significantly improve the network fairly for all road users so as to effectively achieve effectiveness in routing and re-routing in a timely manner. These improvements have a direct bearing on travel time for all road users on average. As a result, fuel consumption improves and green-house emissions reduce. An intelligent routing algorithm or set of algorithms are required to assist drivers so that the selfish nature of driving is mitigated with methodologies that result in benefits for all road users. Methods that benefit all road users would result in both drivers and passengers achieving travel satisfaction and reducing their frustrations. When a shortest route is known between an origin and a destination or a relative origin and destination (i.e. common points in the travel metrics that traffic moving to similar zones or passing through similar zones must transverse), all drivers would gravitate to that route causing the paradox scenario to become true. Traffic will load into this common perceived shortest path and would cause congestion referred to as the price of Anarchy. This situation is known as the Braess Paradox named after a German physicist, Braess who explored the paradox in 1968. From that time, it has been a well-known phenomenon that when drivers have knowledge of a shortest path, they will all elect that path and would increase the cost of travel. For example, when the 42nd street was closed in New York City, instead of the predicted gridlock traffic, flow improved because the road network became more balanced as opposed to the perceived shorter low weight link, which sooner than later became a high cost link due to the increase in traffic which works counter to the belief that traffic flow ought to improve due to its shortness. The Scientific American article [1] [2] highlights a recent example from Seoul, South Korea, in which the inverse/ converse occurred, that is, the removal of a road made users better off.

The proposed algorithms are aiming to increase the expectations of optimized consistent travel times that provide the following: 1) consistent travel times; 2) reducing passengers and drivers frustrations; 3) multiple shortest routes with time travel being as important as distance travelled; 4) re-routing dynamically on the shortest paths when traffic is re-assessed while travelling is taking place; 5) reduced Carbon Dioxide, Carbon Monoxide and Hydrocarbons; 6) reduced unnecessary re-routing.

2. RELATED WORKS

2.1. Inrix

A leading provider of highly accurate traffic services and real time traffic information base developed temporal accuracy for drivers to collect estimated travel times for multiple routes from which the drivers can reassess the routes and their requirements in order to elect the suitable routes.

The danger with this is that the drivers could all elect a common most attractive route to them thereby creating a Braess paradox. Partners for the INRIX project include Ford, TOMTOM, and Mapquest et al. [3] . [4] also discusses self-learning artificial intelligence approaches that learn the pattern of traffic to become predictive. Transportation databases as discussed in [5] can help to be online road side infrastructure to tap in, in organizing traffic and routing information to on-board devices on vehicles.

Waze [6] is a social GPS application for drivers to share traffic information, which recently helped to reduce the highly publicized Los Angeles' so called Carmageddon when the world's busiest road was closed for repairs on July 15th 2011. Garmins POI Loaders determine whether a file contains speed and proximately alert points based on Specific criteria [7] . POI refers to relative Points Out and Points in (POI), typically traffic lights will exhibit in out points, which is a separate cost from over sensitive routing algorithms, intelligently re-organizing and reconfiguration of the network under consideration.

2.2. Related Approaches to Resolve the Shortest Path Methods

There has been a shift in scientific circles to focus on the shortest travel time rather than the geographical shortest route as discussed by [8] in the examination of time based vectors for consistent travel time compared to distance based vectors. Discussions by [9] reveal approaches involving Genetic algorithms to project which route would yield the best travel time between a known origin or input origin "O" and a known destination "D" at a time "t". [10] Also applies Genetic Algorithms with a discussion on parameters that include travel time along with distribution, journey distance and speed. Related to the discussion in [9] and [10] is another focusing on travel times using a Genetic algorithm method in [11] known as "random key", which runs a form of random algorithm to route vehicles. The cost of Genetic Algorithms and the infancy in the research do not make them attainable in many environments. There is also an element of computational overheads about them. Reinforcement learning systems discussed by [12] do a lot to reduce computational overheads by applying a "minimal data" approach, yet allowing some learning and interpretation of the minimum relevant data for dynamic decision making. "Minimal data usage is also discussed and justified by [13] . [14] attempted an approach to reduce computational overheads but their efforts are not holistic as their proposed manipulation based on priority requires some increase in data being handled because vehicles have to poll priority weights hence an increase in data. Their efforts elsewhere to reduce computational overheads in EBkSP compared to RKSP are subdued as a result. The contributors in [14] build on the idea of multiple best routes (k paths) adapted from [15] . They describe three strategies for load balancing and routing. These strategies are 1) Dynamic Shortest Path (DSP); 2) Random balancing considering future Vehicle (RKSP); and 3) Entropy Balanced shortest Path (EBkSP) which balances the network at a relatively least cost.

The section below describes each the three strategies:

2.2.1. Dynamic Short Paths (DSP)

Dynamic shortest path is viewed as a traditional re-routing strategy that elects the shortest path for vehicles. It is the simpler one of the three and it computational cost is the lowest of them and can be described as $O(E + V\log(V))$, where E is the number of road segments and V is the number of intersections it has been found not to be suited to switching high volumes of traffic load. Seems to work efficiently in manageable, highly predictable loads but fails when this threshold of predictability is surpassed as it is then viewed as simply being able to shift congestion from one segment to another that did not have it before so that there is no net benefit in such circumstances. Also the risk of a Braess Paradox situation we explained becomes increasingly high with this strategy.

2.2.2. Random K Shortest Paths (RKSP)

For each vehicle that is to be re-routed, a number of shortest paths (K) are computed i.e. K-shortest paths. The system assigns each selected vehicle to a selected path (one of the K paths) randomly. The objective is to overcome the limitation in the DSP which would switch congestion to another zone. Route balancing is achieved well by this strategy because congestion transfer is avoided. Naturally a higher price to pay for the computational complexity with the cost formulae being adjusted to $O(KV(E + V\mathrm{Log}(V)))$ [15] in "a procedure for computing the k shortest paths", which increases linearly with K. With multiple K paths, concern is often on the variations on travel times between a similar origin O and a particular destination D increasing significantly. Improve versions of the Random K shortest paths have extra computation to ensure that the difference between the shortest travel time on a path Ks and the longest travel time on the model Ki is kept to a value of at most 20% difference.

2.2.3. Entropy Balanced K Shortest Paths (EBKSP)

This method is perceived the better of the three because in addition to possessing the advantages that "Random K Shorter Paths" has over the Dynamic shortest path, it overcomes the computational costs which random K shortest paths has so that information on re-routing etc. Information is pushed to vehicles long before congestion builds but computational time increase the latency involved in communicating.

2.3. Induction Loops (E1) and Detectors (E3)

In SUMO, an E1 is an Induction loop which gives feedback on lane, position on lane, aggregation time and on which vehicle to route. An E3 is a detector that measures congestion by calculating the amount of vehicles on halt, the speed, mean halt times, queue length, and the mean duration of halts within detectors. The E3 is a multi-entry, multi-exit detector [16] Using E1and E3 Induction loops (viii) Intelligent Traffic Lights again with built in E1 and E3 algorithms will assist in making the network intelligent and responsive and to become self-organizing depending on the suitability of the junction as too much of one technology can increase latency and reroute traffic unnecessarily which would be counter to the objective of this project.

3. SYSTEM MODEL

The system builds on optimizing consistent travel times with a smart knowledge base for learning travel patterns and then balancing the load. The intelligence for this is built in the E1 and E3 Induction loops and detectors respectively. E3 detectors are built at intersection points and on the track of highways. Arguably, the logic is centralized at the induction loops or road side units' traffic lights but communicating with drivers on board equipment. Key components are duarouter, a tool that computes shortest paths, dynamic user assignment, and optimal path routing as alternatives or extensions to simple shortest path used to re-route traffic based on calibrated parametres. Other tools include net converter, a tool for producing the map or network on which the simulation is to be evaluated and Traffic Control Intelligence (TraCI) built in traffic lights supporting roadside infrastructure and the rest of SUMO tools (Actually all this in the SUMO environment).

Whereas the "original" origin to destination is known to make the system more dynamic relative new origins are to be computed as the knowledge base learns of the new conditions and updates its current position as its new origin and re-computes the new route based on current conditions. Our models will not focus on distinguishing cars as "urgent" as in Entropy balanced K shortest paths (EBKSP) as this would cause a computational overhead. Whereas EBKSP is praised for reducing Computational overhead over Random K shortest paths

(RKSP) by not moving congestion to another section, by pushing re-routing before vehicles pass the intersection and assigning an urgency weighted value on cars, so that most affected vehicles are given prioritized opportunity, it introduces its own additional constraint or extra requirement of ranking vehicles on a priority scale and using this as a paradigm for re-routing. This approach of prioritizing benefits some drivers at the expense of others. We propose a mutual benefit algorithm which spreads the benefits easily so that the earliest arrival time benefits the driver that started the earliest if their origin and destination were to be the same, and the network remained balanced with the remainder of the algorithms assigning edges (routes) to traffic.

The simulation is implemented using the modified example runner.py Python script (@version $Id: runner.py 14247 2013-07-01 11:56:51Z namdre $) originally, as demonstrated in the tutorial for traffic light control via the TraCI interface. We have modified generate_routefile() and run() functions.

4. AVOIDING UNJUSTIFIED RE-ROUTING

Re-routing would be better achieved if the start and the end of congestion are understood linearly so that false congestions do not cause unnecessary re-routing. We will define false congestion as corresponding to a "chain gun" of vehicles for about 1/2 km or even 1KM on say a 5 km link or any realistic threshold that may be set. In this example, it would imply that 4km of the stretch is free of congestion. If loops or road side units were located say less than 500 meters apart say 200 m into the "chain gun" and 700 m into the chain gun respectively, the induction loop would indicate a false alarm which would invoke re-routing. Equally 1 Kms length of slow vehicles probably due to a slow vehicle in front may result in false re-routing information being passed. Therefore alternative distributed and integrated feedback is required to produce realistic global view alerts which would be less costly and smarter. Without smarter approaches, the net effect would bring about introducing delay where there was little or none of it. So, careful design with rational distances is required in order to realize full benefits. Feedback as stated by [17] and [18] must be beneficial to transport intelligent planning. Contributors in [19] and [20] point out to improved traffic flows as solutions to congestion by involving route guidance agents.

Our model will take care of this by ensuring that congestion is declared only when it is longer than 10% of a link being considered. Only when the queue of jammed vehicles is longer than 10% of the linear distance being considered here referred to as the segment. In the scenario we have given above that should be more than 100 meter in a kilometer stretch for each it to be considered congestion, for 10km links 1km of the link having cars following each other on a bumper will warrant a congestion being delayed. Normally roads network are treated as directed, weighted graphs where roads would correspond to intersections, edges to roads segments and weights to estimated travel times. These are the normal parameters on which load balancing is attempted on a road network.

Greenfields model (14) states that the relationship between estimated road speed Vi and traffic density Ki is linear and Ki is the vehicle per meter on a road segment i, and Kjam and Vf are the traffic jam density and free flaw speed respectively on a road segment i and Ti and Li are the estimated travel times and length for the segment:

$$Vi = Vf\left(I - Ki/Kjam\right) \text{ and } Ti = Li/Vi$$

Ki Kjamis the ratio between current-number-of-vehicles and the max-number-of-vehicles.

5. PROBLEM FORMULATION

The Challenge with the Entropy method discussed [14] has potential feedback overheads because when deriving Priority in ranking the vehicle the controlling roadside detectors incur overheads as all vehicles poll their data to this centralized system which must then assign priority values. In the Fully Balanced-Candidate Path algorithm (FBMkP) which we propose over EBkSP, we overcome the identified computational overhead in a different way from the way [12] and [13] approach it by balancing the network fairly for all vehicles without assigning priorities to certain vehicles as this may advantage those vehicles with higher priorities at the expense of others and equally having to store priority values communicated by every vehicle means keeping too much vehicle data. In our FBMkP, the road side units only read the Destination information as vehicles pass them and the E1s and E3s work together in a responsive fashion which evolves dynamically. The network as a whole is designed to fairly route vehicles across the network. The System also take care of unnecessary and costly re-routing.

Algorithm ideas (flow chart): Fully Balanced Multipath-candidate shortest Paths (FBMkP) algorithm (Pseudo Code).

Procedure main
queuelength := obtainRoadSideUnitQueueLengths()
congestedRoads := detectCongestion(queuelength)

vehiclesDestinedCongestedRoad:=selectCongestionDestinedVehicles(congeste dRoads)
trips=updateTrips(vehiclesDestinedCongestedRoad)
newTrips=findkShortestPaths(trips,vehiclesDestinedCongestedRoad)
reroute(newTrips,vehiclesDestinedCongestedRoad)
End Procedure
procedure reroute
for all vehicle in vehicleDestinedCongestedRoad do
{origin,destination}=getVehicleOD(vehicle)
update_paths=computeKPaths(newPaths,origin,destination)
reroute(vehicle,update_path)
end for
end procedure

In our simulation the Origin being in Kalombo Rd and Destination Kabulonga_ Rd in Lusaka, Zambia as represented by the trip file (<trip id = "2" depart = "1" from = "166752211#6" to = "98888584"/>) in terms of Origin and Destination (O-D).

Duo-router, a SUMO tool, computes the k shortest paths between the origin and the destination.

(duarouter—trip-files   trips.xml—net-file   lusaka.net.xml—output-file result.rou.xml).

6. MODEL

In our model, we simulate 3 scenarios and compare the results;

1) Using the shortest path—Traffic flows from O-D using the shortest path. No rerouting is applied in this case.

2) Using congestion detection mechanism—Traffic flows from O-D but partial re-routing occurs but is restricted in terms being done once. The shortest path gets congested at some points, traffic is rerouted to the next possible/alternative shortest path 3) Using intelligence built in traffic lights on scenario 2 such that the Induction loop for feedback known as E3s continually perform re-routing for every congestion hence giving us a fully balanced re-routing algorithm. Traffic flows are rerouted and TLS is controlled by allowing the route which is more likely to have less traffic to have priority.

In the first Scenario, it's assumed that drivers will always choose to use the shortest path and in this case, the shortest path is O-D via Great East road computed by duarouter as shown in the appendix map to this document. Congestion will build up because every driver will default to the route which they think is the shortest. We run the Simulation for 3000 steps. Results obtain as below Example of the Traffic Control loops at Makishi road and Great East road junctions in Lusaka showing roadside infrastructure and vehicles.

E1 and E3 detectors have been used to detect flows on the network. In our Simulation, E3 detectors, will measure congestion by calculating the amount of vehicles on halt. A threshold is set that assumes congestion being true. If congestion is greater than threshold, congestion is set to true and rerouting is triggered by using E1 detectors to determine which vehicles to reroute. In this way congestion is controlled by rerouting traffic to alternative routes as in partially Balanced and Fully Balanced models. Better CO and CO_2 outputs are presented by the Fully Balanced re-routing method as shown in Figure 1 and Figure 2 respectively. In the Fully Balanced Re-routing method, we have set up more iterations and re-Routing intelligence and this method shows more gain in fuel efficiency as shown in Figure 3. Partial re-routing shows partial improvements indicating every degree of multiple shortest paths has improvements over a single shortest path method through the network. To ensure that traffic is balanced over the network, all points where traffic is likely to build up such as near traffic lights and junctions, have detectors placed there to measure congestion and in this way, traffic is rerouted to the alternative shortest path as computed by Duarouter. Feedback onto the traffic lights is provided by both the E1 and E3 systems to make the traffic lights more responsive in a network relay with the loops and detectors here seen as Road side Infrastructure. This results in an FBMkP algorithm integrated with traffic light control Intelligence (TraCI) to create TLC or simply Traffic Light Intelligence.

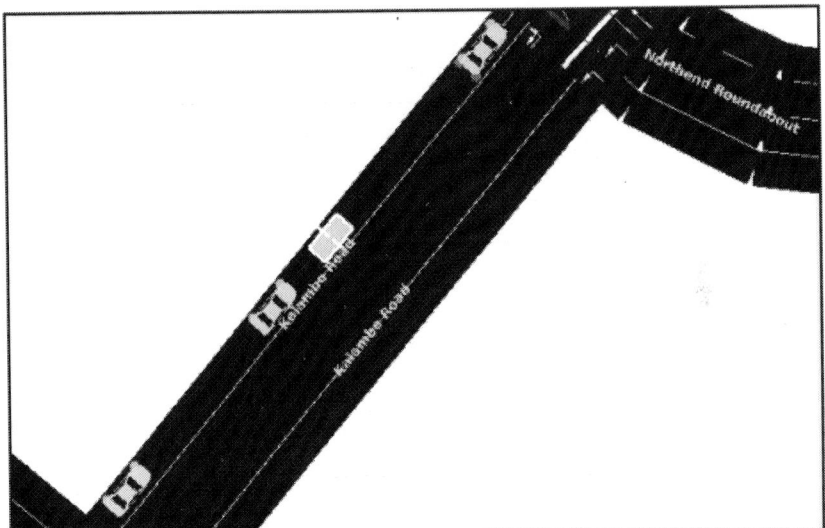

Figure 1. Showing vehicles in the simulation.

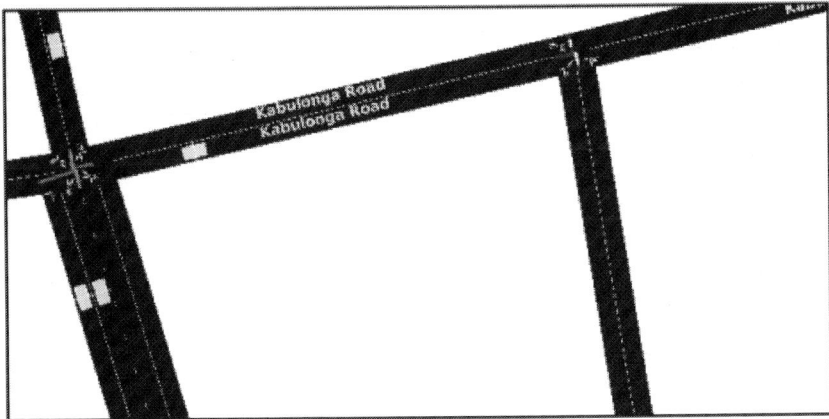

Figure 2. Showing roadside intelligence at an intersection.

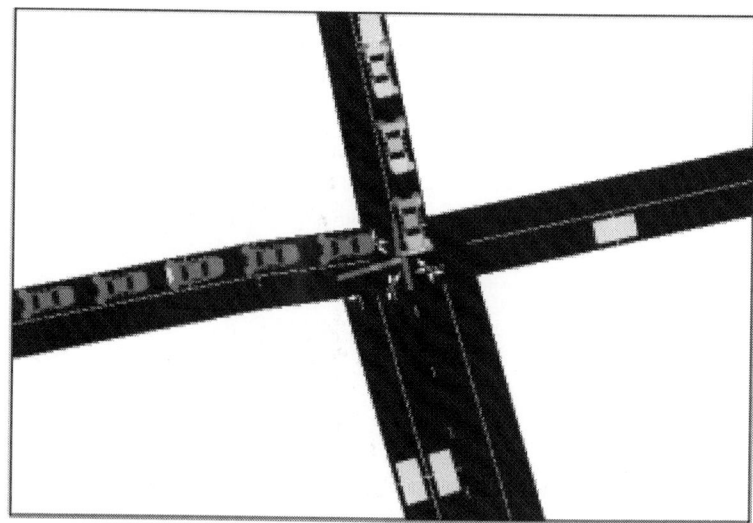

Figure 3. Showing vehicles, roadside units and traffic intelligence as interactive objects.

7. DISCUSSIONS ON THE FIGURES BELOW

In all of the graphs in this paper used, the order of the bars is blue, red, green and purple (this colour, purple is included only where the experiments have included the modified fully balanced algorithm integrated with traffic light control feedback TLC), This is also the order in which the bars are arranged from left to right in Figure 1 to Figure 7 with the exception of Figure 8 where the graph lines vary with time and Figure 9 where they are discussed from bottom to top in the same order. In Figure 5 to Figure 10, an extra examination of an attempt to balance the networks by centring the load balance using traffic light control approach is shown in purple at the very end of the first three strategies is indicated. Included is an additional strategy, the traffic light control (TLC) strategy with balanced re-routing in it and this is shown in purple Bars. What we have been able to demonstrate is motivated by many research groups and requirements in [21] for example, which is a clear testimony of how eager governments are in mitigating green-house effects and costs resulting from longer travel times and fuel costs, health and safety issues, which are directly proportional to vehicular network volumes and average vehicle time spent on the networks. The philosophy detailed in the handbook on emissions factors in road transport [22] , sets out an agenda on the need to quantify emissions on urban transport for better planning for healthy cities. This paper translates its findings along that paradigm as shown in some the figures below relating to emissions.

Some figures also show what we are calling the "Traffic Light Control FBMkP or simply TLC", which is a variation of the Fully Balanced Multiple-Candidates shortest Paths incorporating Traffic Control Intelligence

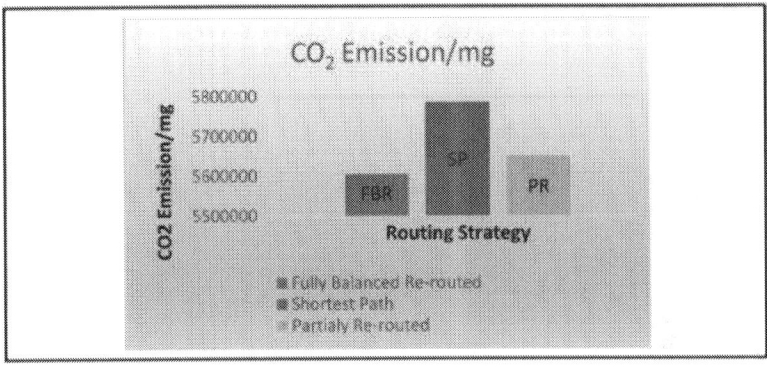

Figure 4. Effect of intelligent re-routing on CO2 emissions.

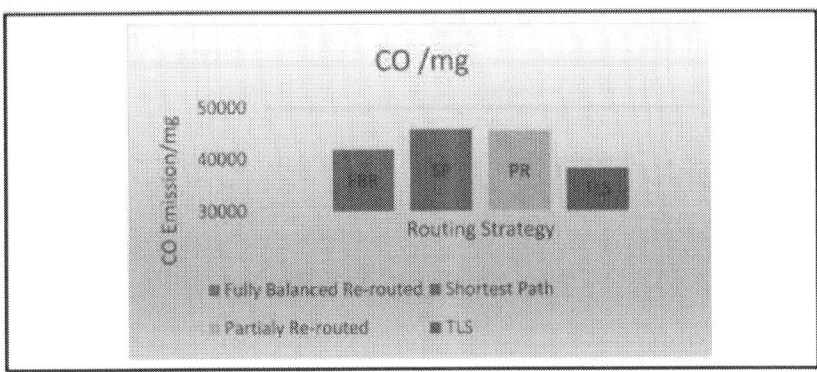

Figure 6. Effect of intelligent re-routing on fuel consumption.

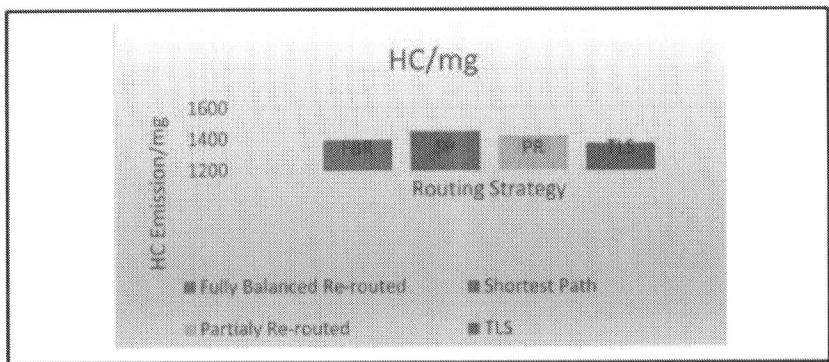

Figure 7. Effect of intelligent re-routing on HCs.

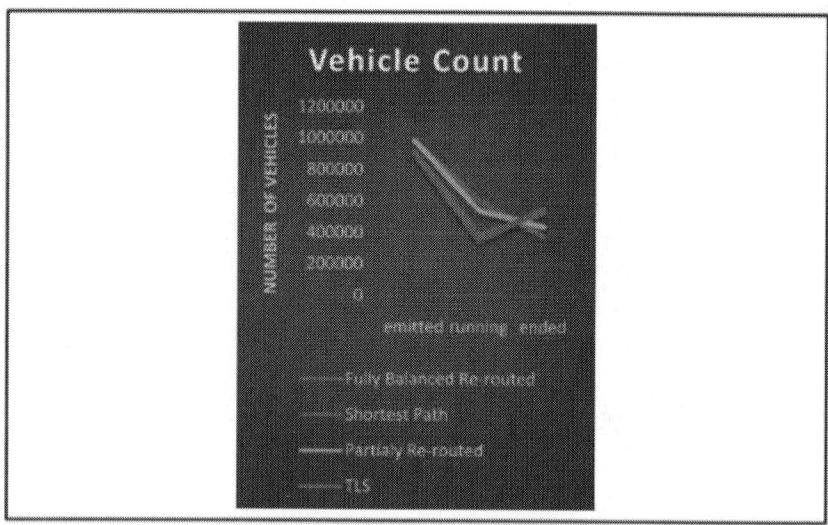

Figure 8. Showing vehicle state in the simulation period.

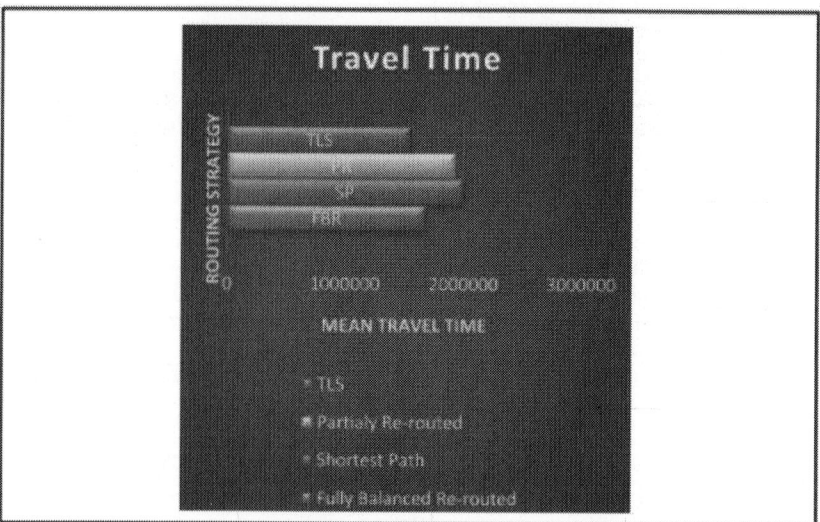

Figure 9. Showing that fully balanced re-routing has the least travel time.

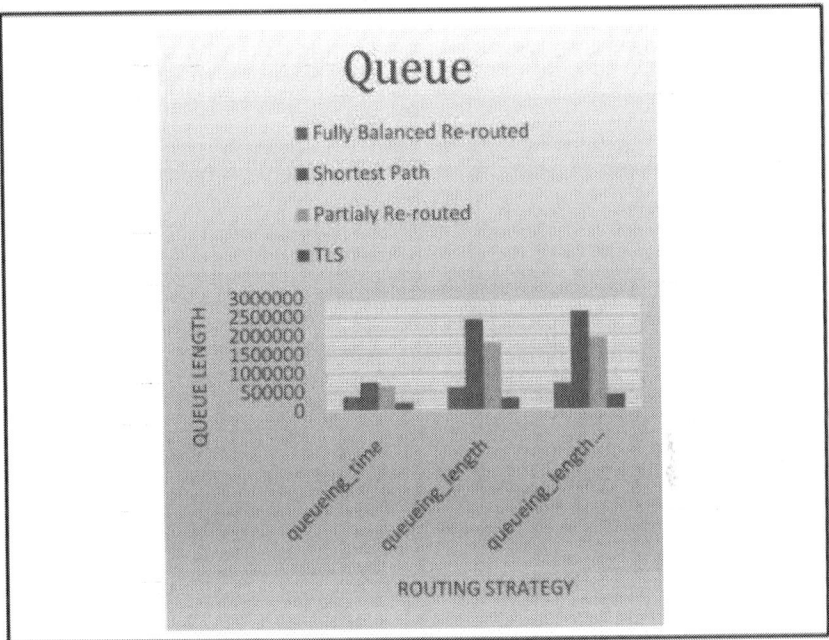

Figure 10. Queuing times and lengths.

(TraCI). TLC is part of our work in progress on achieving a distributed control system between road side infrastructure and traffic light intelligence on improving traffic lights and as such TLC is a proposed component of FBMkP our main algorithm. Our graphs do show that main FBMkP and TLC are yielding best results on a number of measures such as these below.

1) Figure 1 shows us the point on Kalambo road where the simulation begins or where the vehicles are generated from, The SUMO id for this point is "166752211#6".
2) Figure 2 simply shows a typical intersection on Kabulonga road in Lusaka with the road side units for intelligent support.
3) Figure 3 shows vehicles at a traffic light intersection. The lighter coloured vehicles are ready to go as their side is currently on "green". The darker or red cars are in that colour to indicate the state of antsy as they wait for the signal to go.
4) Figure 4 shows us a comparison of results for CO_2 emissions for the routing strategies namely FBMkP (FBR), shortest path only (SP) and Partial Re-routing (PR). The shortest path emits the highest CO_2 and the FBMkP, the least proving that a balanced multiple routing strategy is best for reducing CO_1, CO_2, CO and HC are Hand Book of Emissions Factors (HBEFAs) attributes
5) Figure 5 is a Prototype of the Figure 4 experiments but in it, is carbon monoxide is evaluated against four strategies that now include a variant of

FBMkP known as Traffic Light-control system FBMkP (TLC). TLC performs best in reducing CO. TLC is fully balanced but with extra intelligence on traffic lights to vary the length of "green" or "red" signs on the basis of vehicle queue length.

6) Figure 6 has the same four routing strategies as in 5. The focus is on emitted hydro carbons per mg. FBMkP and FBMkP-TLS (TLS) perform the best. http://www.tdm-beijing.org/ files/news/ Emission% 20 Quantification% 20Workshop/ Schmied_INFRAS_-_Development_of_Emission_Factors_ in_Europe.pdf

7) Figure 7 is discussed in the context of emitted, running and completed vehicles on the road. What we see is that given a sizeable sample of distance, more FBMkP and TLC vehicles completed in a predetermined time. This is an indication of higher predictability on the part of the balanced approaches FBMkP and FBMkP-TLS (TLS) over the others

8) Figure 8 shows us the total travel times between the origin and destination in the experiments. TLS and FBMkP (FBR) obtain the least travel time. As TLS performs slightly better than FBMkP, the factor of traffic light intelligence is also assessed in this way.

9) Figure 9 gives us yet another indication of how the fully balanced algorithms FBMkP and TLC result in the least time on the network between origin and destination. The focus is much on getting the least time and not the least kilometres travelled between the same origin and destination.

10) Figure 10 discusses queuing times and queuing lengths for each of the four strategies as collected from the simulation.

11) Figure S1 in the appendix shows a section of Lusaka, Zambia which is the city used for our simulations in this paper with the shortest path shown in red between the origin and the destination.

The ultimate solution is what we have coined as the "Fully balanced multiple shortest paths" (FBMkP) for short. The FBMkP, unlike the entropy balanced method that assigns priority values to cars, will be fully balanced for all cars giving them an equal opportunity instead of assigning urgency on specific vehicles as this would increase computational overheads. In FBMkP, we have built the intelligence in the induction loops E3s and E1s. Since routes are fewer than cars, the computational overhead is lower when the responsibility is held by road side infrastructure. Instead of all the cars polling their urgency or weighted indices, each car only polls its destination once and loops relay its next edge after every intersection.

8. CONCLUSIONS

From what has been observed and recorded in the experiments, there is a relationship between travel time and emissions. The higher the latency, the greater the carbon footprint. As this is in a multi-valued objective set up, Pareto optimization reveals that a balance between speed and carbon footprint has to be obtained and that speed and time should also be in a balance as too much of any of the two has the potential to increase the carbon footprint for example, where vehicles are in a gridlock shortest path or a state of the braess paradox, more carbon will be emitted and time of travel will be higher, equally high speeds indicate more carbon per kilometre travelled than moderate speeds. The efficiency of the FBMkP algorithm and TLC (a variant of FBMkP) is set up over two main issues namely its simplicity in balancing for all road users rather than for prioritizing some, hence it does not transfer overheads to other segments and secondly in its integration of feedback between vehicles and road side units at several intersections. The systems should also have the intelligence to check the weight of the routes that we are selecting, so that the results are dynamic per time of day and this would mean a certain selection set would not always yield the better results over other paths for the same O-D matrix. See details detailing some results achieved so far. Further assessment of factors in a multi-valued environment using Pareto optimisation and will be investigated using Lusaka, London and New York to build a research result that could be useful to London transport and other stakeholders. The slight increase in fuel consumption in FBMkP-TLC as shown in Figure 6can be attributed to increased speeds in free flow due to the proper balance of TLC which may see an increase in speed due to increased space per vehicle in the network.

REFERENCES

1. Nagurney, A. (2009) Scientific American, The Braess Paradox and the Price of Anarchy "Detours by Design". The Braess Paradox.
2. Translation of the Braess 1968 Article from German to English Appears as the Article "On a Paradox of Traffic pLanning," by Braess, D., Nagurney, A. and Wakolbinger, T., in the Journal Transportation Science, 39, 2005, 446-450.
3. www.INRIX.com
4. Sadek, A. and Basha, N. (2005) Self Learning Intelligent Agents for Dynamic Traffic Routing on Transportation Networks. In: Unifying Themes in Complex Systems, Springer, 503-511.
5. Wolfson, O. and Xu, B. (2010) Spartial-Temporal Databases in Urban Transportation. Bulletin of the IEEE Computer Society Technical Committee on Data Engineering, 33, 1-8.
6. http://gigaom.com/2011/07/05/waze-prepares-for-its-closeup-with-carmageddon/
7. www.garmin.com/products/poiloader
8. Jenelius, E. and Koutsopoulous, H.N. (2012) Impact of Sampling Protocol on Bias and Consistency in Travel Time Estimation on Probe Vehicle Data.http://home.abe.kth.se/~jenelius/JK_2013.pdf
9. Lin, C.-H., Yu, J.-L., et al. (2009) Genetic Algorithm for Shortest Driving Time in Intelligent Transportation Systems. International Journal of Hybrid Technology, 2, 21-28.

10. Shandiz, H.T., et al. (2009) Intelligent Transport System Based on Genetic Algorithms. World Applied Sciences Journal, 6, 908-913. http://www.idosi.org/wasj/wasj6(7)/7.pdf

11. M. Gen and Lin, L. (2006) A New Approach for Shortest Path Routing Problem by Random Key-Based GA. Proceedings of Genetic and Evolutionary Computation, Washington, 8-12 July 2006, 1411-1412

12. Abdullai, B. and Kattas, L. (2003) Reinforcement Learning: Introduction to Theory and Potential for Transport Applications. Canadian Journal of Civil Engineering, 30, 981-991.

13. Georgescu, L., Zeitler, D. and Standridge, C.R. (2012) Intelligent Transportation Systems Real Time Traffic Speed Prediction with Minimal Data. Grand Valley State University, Allendale.

14. Pan, J., Khan, M.A., et al. (2009) Proactive Vehicular Re-routing Strategies for Congestion Avoidance. 2012 IEEE 8th International Conference on Distributed Computing in Sensor Systems (DCOSS), Hangzhou, 16-18 May 2012, 265- 272.

15. Lawler, E.L. (1972) A Procedure of Computing the K Best Solutions to Discrete Optimization Problems and Its Application to the Shortest Path Problem. Management Science, 18, 401-405.

16. Krajzewicz, D., et al. (2013) SUMO-Simulation of Urban Mobility-User Documentation Publications, German Aerospace Centre, Germany. http://sumo-sim.org/userdoc/Publications.html

17. Banks, J. (2002) Introduction to Transportation Engineering. McGraw-Hill, New York.

18. Levingson, D. and Kumar, A. (1994) Integrating Feedback into the Transportation Planning Model: Structure and Application. Transportation Research Record, No. 1413, 70-77.

19. Zhu, Z. and Yang, C. (2011) Visco Elastic Traffic Flow Model. Journal of Advanced Transformation, 47, 635-649.

20. Arnaout, G. and Bowling, S. (2011) Towards Reducing Traffic Congestion Using Cooperative Adaptive Cruise Control on a Freeway with a Ramp. Journal of industrial Engineering and Management, 4, 699-717.

21. Research and Innovative Technology Administration. ITS Strategic Research Plan, 2010-2014. Executive Summary.http://www.its.dot.gov/strat/pdf/ITStrategicresearch/Jan2010.pdf

22. Schmied, M. (2012) Moving towards Emission Quantification in Urban Transport. Handbook on Emissions Factors in Road Transport (HBEFA), Beijing, 23 October 2012.

CHAPTER 6

Adaptive Cascade Generalized Predictive Control

Tao Geng[1], Jin Zhao[2]

[1]Academy of Physics and Electroics, Henan University, Kaifeng, China
[2]Academy of Control Science and Engineering, Huazhong University of
Science and Technology (HUST), Wuhan, China

ABSTRACT

Cascade control is one of the most popular structures for process control as it is
a special architecture for dealing with disturbances. However, the drawbacks of
cascade control are obvious that primary controller and secondary controller
should be tuned together, which influences each other. In this paper, a new
Adaptive Cascade Generalized Predictive Controller (ACGPC) is introduced.
ACGPC is a method issued from GPC and the inner and outer controllers of a
cascade system are replaced by one cascade generalized predictive controller,
where both loops model are updated by Recursive Least Squares method.
Compared with existing methods, the new method is simpler and yet more
effective. It can be directly integrated into commercially available industrial
auto-tuning systems. Some examples are given to illustrate the effectiveness and
robustness of the proposed method.

Keywords: Adaptive Cascade Generalized Predictive Control, Model
Identification, Cascade Control

1. INTRODUCTION

Cascade control is one of the most popular structures for process control as it is
a special architecture for dealing with disturbances. It is widely used in practice,
sometimes in a transparent way (embedded into the electronics of a
servoactuator [1] , controlled power supply[2] , process control [3] etc.). The
general block diagram of cascade control is shown in Figure 1. Cascade control

refers to the design of a control loop for one primary variable by means of multiple sensors and/or actuators and, in its basic configuration; it consists of two cooperative SISO control loops with different time constants. The time constants in inner loop and outer loop are always different. And this difference allows for separate design using conventional techniques.

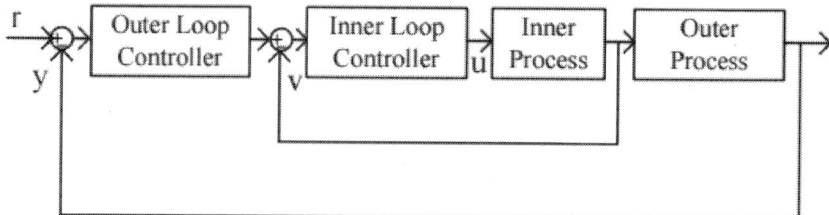

Figure 1. General block diagram of cascade control.

Previous researchers have proposed relay-based auto-tuning techniques to facilitate the design of cascade control systems. The methods proposed by Hang et al. [4] and Vivek and Chidambaram [5] need sequential application of the conventional relay-based auto-tuning approach, and are therefore still time consuming. The sequential tuning procedure has been improved so that only single relay experiment is required for auto-tuning [6] - [9]. However, an off-line or ad hoc experiment must be performed in these methods. For example, Leva and Donida [6] performed test with relay cascaded to an integrator, and Mehta and Majhi [8] restricted the secondary controller to controller during the relay test. Besides the relay-based method, Visioli and Piazzi [10] proposed an automatic tuning method consisting of an open-loop test for cascade control system. Veronesi and Visioli [11] recently proposed simultaneous closed-loop automatic tuning method for cascade controllers. Their method evaluates the set-point step response of cascade control system.

Whatever, based on the PID there are always two controllers needed to be configured. And obviously, the drawbacks of cascade control are obvious that primary controller and secondary controller should be tuned together, which influences each other. If there is not a substantial difference in time constants, although this strategy can still be pursued, the loop design cannot be made independently and based on SISO techniques. So tuning is not intuitive: centralized configurations might be preferable. CGPC [12] [13] is a method issued from GPC and the inner and outer controllers of a cascade system are replaced by one cascade generalized predictive controller. In this control paradigm, there is only one controller configured. If the models of inner process and out process were known, the controller can be auto-configured. This paper proposed an adaptive CGPC cascade controller shown in Figure 2. The both loops model are SISO. And the model can be identified by the classical method, respectively.

2. GPC WITH CONSTRAINTS

GPC adapts the model so-called Controlled auto regressive integrated moving average (CARIMA) model [14] [15].

$$A\left(z^{-1}\right)y = B\left(z^{-1}\right)u + \frac{T\left(z^{-1}\right)}{\Delta}e$$

(1)

where, $T\left(z^{-1}\right)$ is the model of noise, but it is commonplace to treat $T\left(z^{-1}\right)$ as a design parameter. Because it has direct effects on loop sensitivity and so better closed-loop performance will be got with a $T\left(z^{-1}\right)$.

Then, a general form of future predictions is

$$\underline{y} = \Gamma\Delta\underline{u} + \underline{y}_{\text{free}}$$

$$\underline{y}_{\text{free}} = \tilde{P}\Delta\tilde{\underline{u}} + \tilde{Q}\tilde{\underline{y}}$$

(2)

And the followed can be derived

$$\Gamma = C_D^{-1}C_B$$

(3)

where, $D\left(z^{-1}\right) = \Delta A\left(z^{-1}\right)$

$$\begin{cases} \tilde{P} = C_T P - \Gamma H_T \\ \tilde{Q} = C_T Q + H_T \end{cases}$$

(4)

C_T is the toeplitz matrix of $T\left(z^{-1}\right)$ and H_T is the hankel matrix of $T\left(z^{-1}\right)$ [16] . Constraints on process inputs and outputs make the controller and consequently the entire closed-loop, nonlinear.

Use the cost function and optimization

$$J = \sum_{j=1}^{N_y}\left\|r(k+j)-\underline{y}(k+j)\right\|_2^2 W_y(j) + \sum_{j=1}^{N_u}\left\|\Delta\underline{u}(k+j)\right\|_2^2 W_u(j)$$

(5)

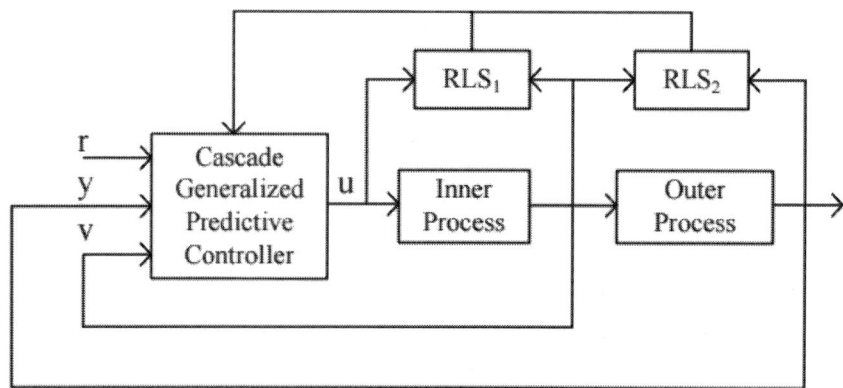

Figure 2. Proposed control paradigm of adaptive cascade control.

where, $r(k+j)$ is j step ahead reference. N_y is receding horizon and N_u is control step. W_y, W_u is the output and input weight factor. Described in matrix form, the GPC with constraints is converted to be the optimization problem which is

$$J = \left\| r - \Gamma \Delta u - y_{free} \right\|_{W_y}^2 + \left\| \Delta u \right\|_{W_u}^2$$

$$\min_{\Delta u} J = \Delta u S \Delta u^T + 2 f^T \Delta u$$

$$\text{s.t. } C\Delta u - d_k \le 0 \tag{6}$$

where, J is subject of the following constraints. Where S is positive definite and f, d_k are time varying (dependent on the current state).

$$S = \Gamma^T W_y \Gamma + W_u, \quad f = -\Gamma^T W_y \left(r - y_{free} \right) \tag{7}$$

The (6) is a standard quadratic programming with constraints. The constraints in GPC will be described as the following.

- The Control Law without Constrains Solve the $\dfrac{\partial J}{\partial u} = 0$

The control law without constrains is derived as

$$\Delta u = -S^{-1} f = \left(\Gamma^T W_y \Gamma + W_u \right)^{-1} \Gamma^T W_y \left(r - y_{free} \right) \tag{8}$$

- Input move constraints

Δu is the lower bounds of input move constraints, and $\Delta \bar{u}$ is the upper bounds of input move constraints.

$$\begin{bmatrix} \Delta\underline{u} \\ \Delta\underline{u} \\ \vdots \\ \Delta\underline{u} \end{bmatrix} \leq \begin{bmatrix} \Delta u_k \\ \Delta u_{k+1} \\ \vdots \\ \Delta u_{k+Nu-1} \end{bmatrix} \leq \begin{bmatrix} \Delta\bar{u} \\ \Delta\bar{u} \\ \vdots \\ \Delta\bar{u} \end{bmatrix}$$

Which can be described in vector form as (9).

$$\Delta\underline{U} \leq \Delta\underline{u} \leq \Delta\bar{U} \tag{9}$$

And satisfy the matrix inequality (10)

$$\begin{bmatrix} I \\ -I \end{bmatrix}\Delta\underline{u} - \begin{bmatrix} \Delta\bar{U} \\ -\Delta\underline{U} \end{bmatrix} \leq 0 \tag{10}$$

- Ÿ Input constraints

$\underline{u}$ is the lower bounds of input constraints, and $\bar{u}$ is the upper bounds of input constraints.

$$\underline{u} = C_{I/\Delta}\Delta\underline{u} + \begin{bmatrix} I \\ I \\ \vdots \\ I \end{bmatrix} u_{k-1}$$

where,

$$C_{I/\Delta} = \begin{bmatrix} I & 0 & \cdots & 0 \\ I & I & 0 & \cdots \\ \vdots & \ddots & \ddots & \ddots \\ I & \cdots & I & I \end{bmatrix}$$

$$\begin{bmatrix} \underline{u} \\ \underline{u} \\ \vdots \\ \underline{u} \end{bmatrix} \leq \begin{bmatrix} u_k \\ u_{k+1} \\ \vdots \\ u_{k+Nu-1} \end{bmatrix} \leq \begin{bmatrix} \bar{u} \\ \bar{u} \\ \vdots \\ \bar{u} \end{bmatrix}$$

$$\underline{U} \leq C_{I/\Delta}\Delta\underline{u} + L u_{k-1} \leq \bar{U} \tag{11}$$

The corresponding linear inequalities are

$$\begin{bmatrix} C_{I/\Delta} \\ -C_{I/\Delta} \end{bmatrix} \underset{\rightarrow}{\Delta u} - \begin{bmatrix} \bar{U} - Lu_{k-1} \\ -\underline{U} + Lu_{k-1} \end{bmatrix} \le 0$$

- Output constrains

The output of plants always needs to be constrained in the bounds, as demand of process requirements. And they always are treated as soft constraints. $\underline{Y}$ is the lower bounds of output constraints, and $\bar{Y}$ is the upper bounds of output constraints.

$$\underline{Y} \le y \le \bar{Y}$$

$$\underline{Y} \le \Gamma \Delta \underset{\rightarrow}{u} + y_{\text{free}} \le \bar{Y}$$

$$\begin{bmatrix} \Gamma \\ -\Gamma \end{bmatrix} \Delta \underset{\rightarrow}{u} - \begin{bmatrix} \bar{Y} - y_{\text{free}} \\ -\underline{Y} + y_{\text{free}} \end{bmatrix} \le 0 \tag{12}$$

If all constraints are satisfied, the (6) can be described as

$$C = \begin{bmatrix} I \\ -I \\ C_{I/\Delta} \\ -C_{I/\Delta} \\ \Gamma \\ -\Gamma \end{bmatrix}, \quad d_k = \begin{bmatrix} \Delta \bar{U} \\ -\Delta \underline{U} \\ \bar{U} - Lu_{k-1} \\ -\underline{U} + Lu_{k-1} \\ \bar{Y} - \tilde{P}\Delta \tilde{u} - \tilde{Q}\tilde{y} \\ -\underline{Y} + \tilde{P}\Delta \tilde{u} + \tilde{Q}\tilde{y} \end{bmatrix} \tag{13}$$

qpOASES (quadratic program Online Active SEt Strategy) is an open-source implementation of the recently proposed online active set strategy, which was inspired by important observations from the field of parametric quadratic programming [16] [17] . The standard form is

$$\min_{\Delta \underset{\rightarrow}{u}} J = \Delta \underset{\rightarrow}{u}^{\text{T}} S \Delta \underset{\rightarrow}{u} + 2\Delta \underset{\rightarrow}{u}^{\text{T}} f$$

$$\text{s.t. } lb \le \Delta \underset{\rightarrow}{u} \le ub$$

$$lbA \le A\Delta \underset{\rightarrow}{u} \le ubA$$

The (6) can be converted to be in the qpOASES form. And the followed can be derived from (9), (11), (12)

$$\underbrace{\Delta U}_{lb} \le \Delta u \le \underbrace{\Delta \bar{U}}_{ub}$$

$$\left(U - Lu_{k-1} \right) \le C_{I/\Delta} \Delta u \le \left(\bar{U} - Lu_{k-1} \right)$$

$$\left(Y - \tilde{P}\Delta \tilde{u} - \tilde{Q}\tilde{y} \right) \le \Gamma \Delta u \le \left(\bar{Y} - \tilde{P}\Delta \tilde{u} - \tilde{Q}\tilde{y} \right)$$

And we can get

$$\underbrace{\begin{bmatrix} U - Lu_{k-1} \\ Y - \tilde{P}\Delta\tilde{u} - \tilde{Q}\tilde{y} \end{bmatrix}}_{lbA} \le \underbrace{\begin{bmatrix} C_{I/\Delta} \\ \Gamma \end{bmatrix}}_{A} \Delta u \le \underbrace{\begin{bmatrix} \bar{U} - Lu_{k-1} \\ \bar{Y} - \tilde{P}\Delta\tilde{u} - \tilde{Q}\tilde{y} \end{bmatrix}}_{ubA}$$

Algorithm 1-GPC with constraints

1) Firstly, we can identify the plant $A\left(z^{-1}\right)$, $B\left(z^{-1}\right)$, and Γ, $\tilde{P}$, $\tilde{Q}$ can be calculated from (3), (4), Specify the factor W_y, W_u, $T\left(z^{-1}\right)$, receding horizon N_y, control step N_u, bound of input U, $\bar{U}$, bound of input rate ΔU, $\Delta \bar{U}$, bound of output Y, $\bar{Y}$, calculate the lb, ub, A, initialize QProblem object (qpOASES).

2) sample the output, update $y \leftarrow$.

3) update lbA, ubA and QProblem object, return the optimization value $\Delta u \leftarrow$.

4) update $\Delta u \leftarrow$, output Δu_k .
5) go to (2).

3. CASCADED GENERALIZED PREDICTIVE CONTROL

As shown in Figure 2, Inner process model is described as

$$A_1\left(z^{-1}\right)v = B_1\left(z^{-1}\right)u + \frac{T_1\left(z^{-1}\right)}{\Delta}e_1 \tag{14}$$

And outer process model is described as

$$A_2\left(z^{-1}\right)y = B_2\left(z^{-1}\right)v + \frac{T_2\left(z^{-1}\right)}{\Delta}e_2$$

(15)

Similar to Formula (1) and (2), future predictions for Formula (14) is

$$\Delta \underline{y} = \Gamma_1 \Delta \underline{u} + \tilde{P}_1 \Delta \underline{u} + \tilde{Q}_1 \Delta \underline{\tilde{v}}$$

(16)

$\Box$ means filter by T_1

$$\Gamma_1 = C_{A_1}^{-1} C_{B_1}, P_1 = C_{A_1}^{-1} H_{B_1}, Q_1 = -C_{A_1}^{-1} H_{A_1}$$

$$\begin{cases} \tilde{P}_1 = C_{T_1} P_1 - \Gamma_1 H_{T_1} \\ \tilde{Q}_1 = C_{T_1} Q_1 + H_{T_1} \end{cases}$$

And the future predictions for Formula (15) is

$$\underline{y} = \Gamma_2 \Delta \underline{v} + \tilde{P}_2 \Delta \underline{\dot{v}} + \tilde{Q}_2 \underline{\dot{y}}$$

(17)

$\overset{\bullet}{\Box}$ means filter by T_2

$$\Gamma_2 = C_{D_2}^{-1} C_{B_2}, P_2 = C_{D_2}^{-1} H_{B_2}, Q_2 = -C_{D_2}^{-1} H_{D_2}$$

$$\begin{cases} \tilde{P}_2 = C_{T_2} P_2 - \Gamma_2 H_{T_2} \\ \tilde{Q}_2 = C_{T_2} Q_2 + H_{T_2} \end{cases}$$

By substitution of Formula (16) into Formula (17)

$$\underline{y} = \Gamma_2\left(\Gamma_1 \Delta \underline{u} + \tilde{P}_1 \Delta \underline{\tilde{u}} + \tilde{Q}_1 \Delta \underline{\tilde{v}}\right) + \tilde{P}_2 \Delta \underline{\dot{v}} + \tilde{Q}_2 \underline{\dot{y}} = \Gamma_2 \Gamma_1 \Delta \underline{u} + \Gamma_2 \tilde{P}_1 \Delta \underline{\tilde{u}} + \Gamma_2 \tilde{Q}_1 \Delta \underline{\tilde{v}} + \tilde{P}_2 \Delta \underline{\dot{v}} + \tilde{Q}_2 \underline{\dot{y}} = \Gamma \Delta \underline{u} + y_{\text{free}}$$

where, $\Gamma = \Gamma_2 \Gamma_1$, $y_{\text{free}} = \Gamma_2 \tilde{P}_1 \Delta \underline{\tilde{u}} + \Gamma_2 \tilde{Q}_1 \underline{\tilde{v}} + \tilde{P}_2 \Delta \underline{\dot{v}} + \tilde{Q}_2 \underline{\dot{y}}$

The control law without constrains corresponds with (8).

The control law with constrains for intermediate output

$$\underline{v} = \Gamma_3 \Delta \underline{u} + \tilde{P}_3 \Delta \underline{\tilde{u}} + \tilde{Q}_3 \underline{\tilde{v}} = \Gamma_3 \Delta \underline{u} + v_{\text{free}}$$

$$v_{\text{free}} = \tilde{P}_3 \Delta \underline{\tilde{u}} + \tilde{Q}_3 \Delta \underline{\tilde{v}}$$

(18)

$\Box$ means filter by T_1

$$\Gamma_1 = C_{D_1}^{-1} C_{B_1}, P_1 = C_{D_1}^{-1} H_{B_1}, Q_1 = -C_{D_1}^{-1} H_{D_1}$$

$$\begin{cases} \tilde{P}_3 = C_{T_1} P_3 - \Gamma_3 H_{T_1} \\ \tilde{Q}_3 = C_{T_1} Q_3 + H_{T_1} \end{cases}$$

$$\underline{V} \le v \le \bar{V} \tag{19}$$

$\underline{V}$ is the lower bounds of intermediate output v constraints, and $\bar{V}$ is the upper bounds. By substitution of Formula (18) into Formula (19)

$$\underline{V} \le \Gamma_3 \Delta \underline{u} + \tilde{P}_3 \Delta \tilde{u} + \tilde{Q}_3 \tilde{v} \le \bar{V}$$

The corresponding linear inequalities are:

$$\begin{bmatrix} \Gamma_3 \\ -\Gamma_3 \end{bmatrix} \Delta \underline{u} - \begin{bmatrix} \bar{V} - \tilde{P}_3 \Delta \tilde{u} - \tilde{Q}_3 \tilde{v} \\ -\underline{V} + \tilde{P}_3 \Delta \tilde{u} + \tilde{Q}_3 \tilde{v} \end{bmatrix} \le 0$$

$$C \Delta \underline{u} - d_k \le 0$$

where,

$$C = \begin{bmatrix} I \\ -I \\ C_{I/\Delta} \\ -C_{I/\Delta} \\ \Gamma_3 \\ -\Gamma_3 \end{bmatrix} \qquad d_k = \begin{bmatrix} \Delta \bar{U} \\ -\Delta \underline{U} \\ \bar{U} - L u_{k-1} \\ -\underline{U} + L u_{k-1} \\ \bar{V} - v_{\text{free}} \\ -\underline{V} + v_{\text{free}} \end{bmatrix}$$

The control law with constrains is converted to be standard quadratic programming with constraints problem and can be solved by Algorithm 1.

4. CLASSICAL RLS METHOD

The SISO system can be identified by classical RLS method, which is described as followed.

$$\tilde{y}_k = \theta^T \varphi + e_k$$

$$\theta = \begin{bmatrix} a_1 & \cdots & a_n & b_1 & \cdots & b_m \end{bmatrix}^T$$

$$\varphi = \begin{bmatrix} \tilde{y}_{k-1} & \cdots & \tilde{y}_{k-n} & \tilde{u}_{k-1} & \cdots & \tilde{u}_{k-m} \end{bmatrix}^T \tag{20}$$

$\Box$ means filter by T in Equation (1). where θ is a vector of adjustable model parameters and e_k is the corresponding error at time k. The aim is to select θ so that overall modeling error is minimized. The classical recursive least square algorithm is

$$P_k = \left[1 - \varphi\left[\varphi^T P_{k-1}\varphi + \mu\right]^{-1}\varphi^T\right]P_{k-1}$$

$$\hat{\theta}_k = \hat{\theta}_{k-1} + K(\tilde{y}_k - \hat{\theta}_{k-1}\varphi^T)$$

$$K = P_k\varphi/\mu$$

(21)

In the RLS algorithm, the P_k update equation Equation (21) is very sensitive to the truncation errors and there is no guarantee that P_k will always be positive and symmetric defined. The parameter identified by RLS will have biases from the true parameters, the stability and robustness of the algorithm is very poor. And then P_k is revised as

$$P_k = \left(P_k + P_k^T\right)/2$$

5. VERIFICATION ON PROPOSED SCHEME

Two examples are presented here to illustrate the effectiveness of the proposed tuning method for cascade control systems shown in Figure 2. The parameters of the ACGPC are in the Table1

In simulation, there is an identification process at first 100 s. Both inner and an outer model are identified by the classical RLS, respectively. The final value of model identified by the RLS is shown in the Table 2 and Table 3, respectively. And the Figures 3-6 show model parameters estimated by RLS. There is rarely fluctuation in convergence of parameter estimated. Based on the model, the CGPC is applied to control the cascade system. The outer process outputs (primary outputs) of the control loops are presented in Figures 5-8. The figure clearly shows the CGPC have excellent transient performance. This implies that all the good properties of the cascade control are kept in the ACGPC. This behavior is specific to the cascade structures. And the ACGPC can autotune the CGPC.

Table 2. Identification result of RLS for the real model.

Symbol	Value
Inner loop model A_1	$[1.0000 \quad -0.9048]$
Inner loop model B_1	$[0.0952]$
Outer loop model A_2	$[1.0000 \quad -0.9458]$
Outer loop model B_2	$[0.03]$

Table 3. Identification result of RLS for the real model.

Symbol		Value
Inner loop model	A_1	$[1.0000 \quad -0.8187]$
Inner loop model	B_1	$[0.1813]$
Outer loop model	A_2	$[1.0000 \quad -0.9625]$
Outer loop model	B_2	$[0.0202]$

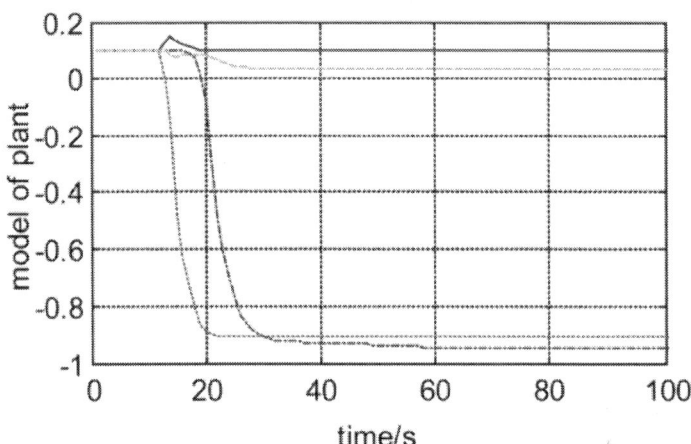

Figure 3. Identification result of RLS for the real model and G1(S).

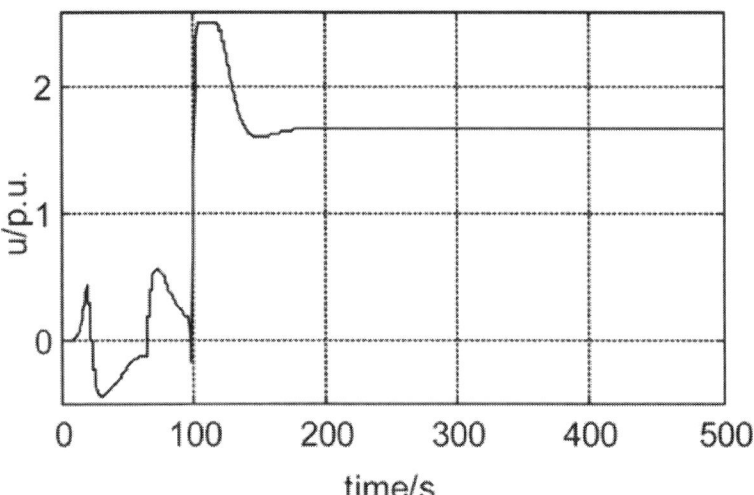

Figure 4. The control signals with and with control input constraints.

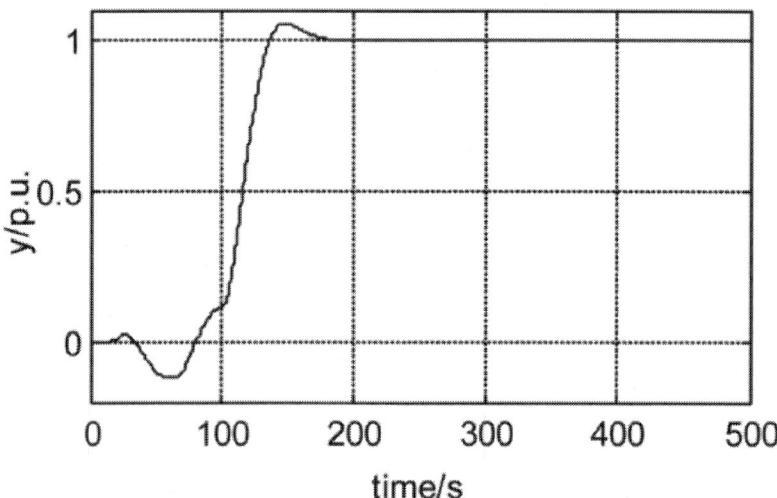

Figure 5. The tracking and regulation performance of the CGPC.

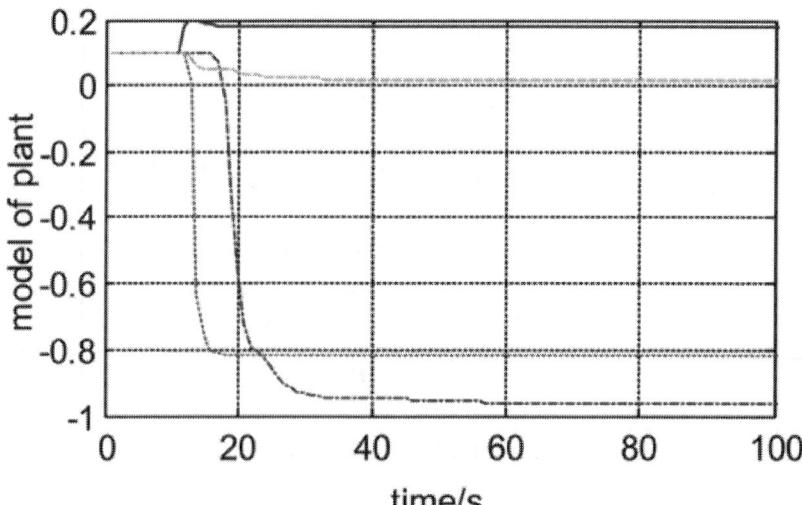

Figure 6. Regulation performance of the CGPC under inner disturbance disturbance which zoomed in Figure 5.

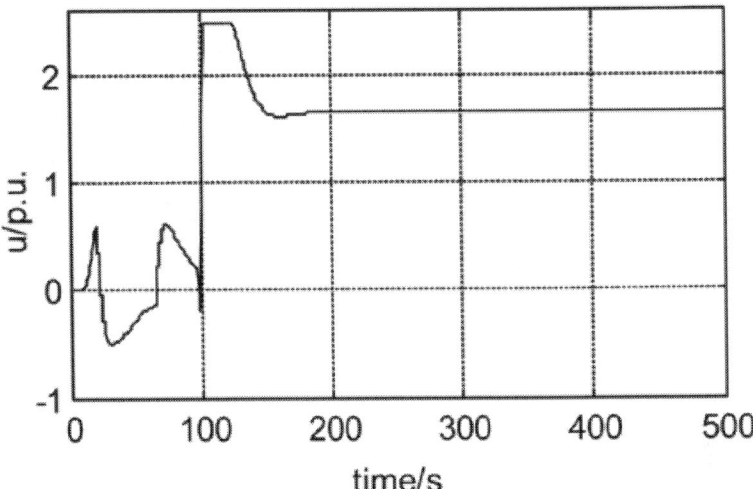

Figure 7. The control signals with and with control input constraints.

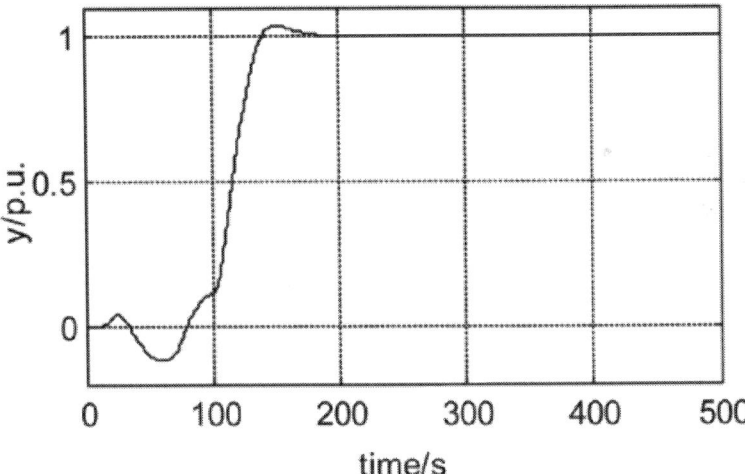

- **Figure 8.** The control signals with and with control input constraints.

Example 1
The inner process is:

$$G_1(s) = \frac{1}{10s+1}$$

The outer process is:

$$G_2(s) = \frac{0.6}{20s+1}$$

- **Example 2**

The inner process is:

$$G_1(s) = \frac{1}{5s+1}$$

The outer process is:

$$G_2(s) = \frac{0.6}{30s+1}$$

6. CONCLUSION

This paper developed an ACGPC method for the cascade control system, which gives the possibility to identify and control some different variables together. Both inner loop and outer loop process model, parameters can be identified using classical RLS method. Consequently, well-identified model based on CGPC can be applied to cascade control system. Finally, two examples were given to show the effectiveness of the proposed method. The method is very straightforward and has been integrated into an existing auto-tuning system. It is now being tested in an electrical drives system and the field results will be reported soon.

ACKNOWLEDGEMENTS

This work was supported by Henan University Science Foundation 2013YBZR013 and National Natural Science Foundation of China with grant number 61273174.

REFERENCES

1. Pisano, A., Davila, A., Fridman, L., et al. (2008) Cascade Control of PM DC Drives via Second-Order Sliding-Mode Technique. IEEE Transactions on Industrial Electronics, 55, 3846-3854.

2. Tsang, K.M. and Chan, W.L. (2008) Non-Linear Cascade Control of DC/DC Buck Converter. Electric Power Components and Systems, 36, 977-989.

3. Wolff Erik, A. and Skogestad, S. (1996) Temperature Cascade Control of Distillation Columns. Industrial & Engineering Chemistry Research, 35, 475-484.

4. Hang, C.C., Loh, A.P. and Vasnani, V.U. (1994) Relay Feedback Auto-Tuning of Cascade Controllers. IEEE Transactions on Control Systems Technology, 2, 42-45.

5. Vivek, S. and Chidambaram, M. (2004) Cascade Controller Tuning by Relay Auto Tune Method. Journal of the Indian Institute of Science, 84, 89-97.

6. Leva, A. and Donida, F.F. (2009) Autotuning in Cascaded Systems Based on a Single Relay Experiment. Journal of Process Control, 19, 896-905.

7. Song, S., Cai, W. and Wang, Y.G. (2003) Auto-Tuning of Cascade Control Systems. ISA Transactions, 42, 62-72.

8. Tan, K.K., Lee, T.H. and Ferdous, R. (2000) Simultaneous Online Automatic Tuning of Cascade Control for Open Loop Stable Processes. ISA Transactions, 39, 233-242.

9. Visioli, A. and Piazzi, A. (2006) An Automatic Tuning Method for Cascade Control Systems. Proceedings of the IEEE International Conference on Control Applications, Munich, July 2006, 2968-2973.

10. Veronesi, M. and Visioli, A. (2011) Simultaneous Closed-Loop Automatic Tuning Method for Cascade Controllers. Control Theory and Applications, 5, 263-270.

11. Benyó, I. (2006) Cascade Generalized Predictive Control—Applications in Power Plant Control. University of Oulu, Oulu.

12. Benyó, I. (2007) Cascade Generalized Predictive Controller: Two in One. International Journal of Control, 79, 866-876.

13. Clarke, D.W., Mohtadi, C. and Tuffs, P.S. (1987) Generalized Predictive Control—Part II. Extensions and Interpretations. Automatica, 23, 149-160.

14. Clarke, D.W., Mohtadi, C. and Tuffs, P.S. (1987) Generalized Predictive Control—Part I. The Basic Algorithm. Automatica, 23, 137-148.

15. Rossiter, J.A. (2003) Model-Based Predictive Control-A Practical Approach. CRC Press, Washington DC, 54-58.

16. Ferreau, H.J. (2009) qpOASES (Quadratic Program Online Active SEt Strategy) Version 2.0.

17. Ferreau, H.J., Bock, H.G. and Diehl, M. (2006) An Online Active Set Strategy for Fast Parametric Quadratic Programming in MPC Applications. Proceedings of the IFAC Workshop on Nonlinear Model Predictive Control for Fast Systems, Grenoble, 2006, 21-30.

CHAPTER 7

Handling model uncertainty in model predictive control for energy efficient buildings

M. Maasoumya,*, M. Razmarab, M. Shahbakhtib, A. Sangiovanni Vincentellic

[a]Mechanical Engineering Department, University of California, Berkeley, CA 94703, United States
[b]Department of Mechanical Engineering–Engineering Mechanics, Michigan Technological University, Houghton, MI 49931, United States
[c]Department of Electrical Engineering and Computer Science, University of California, Berkeley, CA 94703, United States

ABSTRACT

Model uncertainty is a significant challenge to more widespread use of model predictive controllers (MPC) for optimizing building energy consumption. This paper presents two methodologies to handle model uncertainty for building MPC. First, we propose a modeling framework for online estimation of states and unknown parameters leading to a parameter-adaptive building (PAB) model. Second, we propose a robust model predictive control (RMPC) formulation to make a building controller robust to model uncertainties. The results from these two approaches are compared with those from a nominal MPC and a common building rule based control (RBC). The results are then used to develop a methodology for selecting a controller type (i.e. RMPC, MPC, or RBC) as a function of building model uncertainty. RMPC is found to be the superior controller for the cases with an intermediate level of model uncertainty (30–67%), while the nominal MPC is preferred for the cases with a low level of model uncertainty (0–30%). Further, a common RBC outperforms MPC or RMPC if the model uncertainty goes beyond a certain threshold (e.g. 67%).

Keywords: Modeling uncertainty; Building control; Model predictive control; Building HVAC system; Kalman filter

1. INTRODUCTION

Reducing the energy consumption of buildings by designing smart controllers for operating the HVAC system in a more efficient way is critically important to address energy and environmental concerns [1]. Advanced control algorithms are considered major enablers to achieve higher energy efficiency in commercial buildings. Entire sections of the ASHRAE 90.1 standard [2] are dedicated to the specification of control requirements. Although the optimal control of an HVAC system is a complex multi-variable problem, it is standard practice to rely on simple control strategies that include on–off controllers with hysteresis, and PID controllers.

For optimal control design a thermal model of the building is needed. To achieve building-level energy-optimality, building model should be able to capture the interaction between physically connected spaces in the building, heat storage in walls, and provide an accurate prediction of temperature in the building. Control algorithm on the other hand, should be able to minimize energy consumption and optimize thermal comfort by exploiting occupancy schedules, weather forecast, and system dynamics (i.e. a model to predict temperature evolution of indoor air), and satisfy state (i.e. room air temperature and wall temperatures) and inputs (i.e. discharge air temperature and air mass flow rate) constraints and operate the HVAC system of the building in an optimal fashion within the range of operation of the components.

Model predictive control (MPC) is a promising control strategy that is capable of addressing all the aforementioned criteria and has shown results for achieving higher energy efficiency in buildings [3], [4], [5], [6] and [7]. MPC can provide a potential building energy saving of 16–41% compared to the commonly used rule-based building HVAC controllers [3], [8] and [9]. Other advantages of MPC for building HVAC systems include robustness, tunability, and flexibility [3]. Application of MPC for building energy control has been reported in the literature [3], [8], [7], [10], [11], [12], [13] and [14]. There are different variations of nominal MPC such as distributed [13] and [15], robust [10] and [16]and stochastic [7] and [4]. MPC strategies to systematically address various challenges in building energy control. In [17] the authors propose a computationally tractable approximation of the nonlinear optimal control problem by which they optimize the predicted mean vote (PMV) index, as opposed to the static temperature range. A robust control strategy based on static pressure and supply air temperature reset control is presented in [18] for variable air volume (VAV) system. [19] proposes a controller based on a three mode robust control strategy where each mode addresses different control objectives and conditions; this proposed controller is robust in different load conditions. Authors of [20] showed that in presence of model uncertainty an H_∞-robust controller achieves not only a robust performance on set-point tracking of the air-handling unit but also less energy consumption compared to the pole-placement controller. Authors in [21]observed that indoor zone volume acts as system's bifurcation parameter. A multi-variable regulation strategy based on feedback linearization is used to prevent secondary Hopf bifurcation. The designed control improves the limit cycle behavior and decreases indoor temperature variation.

However, these control techniques rely heavily on a perfect (or almost perfect) mathematical model of the building and a perfect estimation of the unmodeled dynamics of the system [3] to achieve considerable energy saving. In [8] the authors argue that, based on industrial experience, modeling is the most time-demanding and costly part of the automation process. Recently, numerous mathematical models of building thermal dynamics have been proposed in the literature. Resistor–capacitor (RC) models with disturbances to capture unmodeled dynamics have been proposed in [9], [3] and [22]. A bilinear version of an RC model is presented in [7] that takes into account weather predictions to increase building energy efficiency. In [23], the authors found that time varying properties such as occupancy can significantly change the dynamic thermal model and influence how building models are identified. While modeling a multi-zone building, the authors of [23] observed that the experimental data often did not have sufficient quality for system identification and hence, proposed a closed-loop architecture for active system identification using prediction-error identification method (PEM). Although a great deal of progress has been made in modeling the thermal behavior of building envelope and HVAC system [9], [22], [3], [24], [5] and [6], the random nature of some components of these systems makes it very hard to predict, with high fidelity, the temperature evolution of the building using mathematical models. Buildings are dynamical systems with uncertain and time-varying physical and occupancy characteristics. The heat transfer characteristics of a building are highly dependent on the ambient conditions. For instance, heat transfer properties such as convective heat transfer coefficient h, of peripheral walls is dependent on outside temperature, wind speed and direction. Also, unmodeled dynamics of a building [3] is function of (1) *external factors*: ambient weather conditions such as radiative heat flux into the walls and windows, and cloudiness of the sky, and (2) *internal factors:* such as occupancy level, internal heat generation from lighting, and computers. These quantities are highly time-varying and therefore the dynamics of the building and, consequently, parameters of the mathematical model need to constantly adapt to this change over time.

One approach to increase the accuracy of the linear building models is to use an adaptive parameter estimation technique such that the building parameters are updated as the environment changes which leads to an adaptive modeling framework. Although this technique has been used for joint state-parameter estimation in other applications [25],[26] and [27], to the best of the authors' knowledge, this paper is the first study on developing adaptive modeling framework for simultaneous estimation of building*parameters*, *states* and *unmodeled dynamics*.

Four approaches can be taken to model dynamic behavior of buildings and overcome model uncertainty for building controls:

1. Develop detailed nonlinear physical models for building [28] and [29], and infer time-varying factors such as weather conditions, occupancy level, etc. [30].
2. Incorporate sensors to measure time-varying factors [31] and [32].

3. Develop an adaptive computationally efficient model which learns and updates building time-varying parameters.
4. Design building controllers which are robust to model uncertainties.

The first approach is typically computationally expensive. Consequently, its application for real-time building controls is limited. The second approach provides accurate information about time-variation of influential factors on building performance but this approach is not cost-efficient and can be limited by the possibility of adding new sensors to a building. The third and fourth approaches are promising and they are the focus of this paper. In particular, we develop a parameter adaptive building (PAB) model and design a robust MPC for buildings. In this paper we build upon our previous work reported in [3],[9], [33], [22] and [16].

The overall contribution of this paper is putting together modeling, control and co-design in a coherent framework to develop a methodology for selecting a controller type (i.e. RMPC, MPC, or RBC) as a function of building model uncertainty. Particular contributions are:

1. A novel adaptive modeling framework for building predictive control is presented. The modeling framework also illustrates the application of unscented Kalman filter (UKF) technique for building online parameter identification and state estimation
2. Impact of model uncertainty on HVAC predictive controllers is characterized.
3. A new RMPC structure that uses disturbance feedback parameterization of the input is introduced. We show that this parameterization reduces the number of decision variables of the optimization problem and hence results in a faster alternative to the existing parameterizations in the literature, while maintaining the performance level of the RMPC.
4. A guideline for choosing an MPC versus an RMPC, versus a rule-based control depending on the level of model uncertainty is proposed.

The rest of the paper is organized as follows. Section 2 explains the experimental setup used to collect data for this study. We present the proposed parameter adaptive building (PAB) model and the developed parameter/state estimation technique in Section 3. Controller design and performance results for MPC and RMPC, as well as the indices based on which we assess the performances of the introduced controllers are presented in Section 4. Conclusions are drawn in Section 5.

3. TEST-BED AND HISTORICAL DATA

The model studied in this paper is a model for an office room in Lakeshore building at Michigan Technological University. This room is surrounded by two rooms and a corridor in the building and connected to the outdoor area with a thick concrete wall and two south-oriented double-layered windows. Each room is equipped with temperature and humidity sensors (Uni-curve Type II) with the temperature accuracy of ± 0.2°C as part of the building management system (BMS). We have used a different sensing device, (temperature data logger with accuracy of ±0.8 °C) to account for spatial temperature variation in the room and sensing accuracies of individual sensors. Location of the zone sensors are shown in Fig. 1. Temperature readings from these two sensors are shown inFig. 2. We follow the methodology proposed in [33] to find the temperature measurement accuracy, which is obtained to be ±0.8 °C, and is used in the state estimation algorithm which is described in Section 3.3. Outdoor temperature is also measured by the BMS system.

Figure. 1. Location of the temperature sensors in the test-bed. The sensor 1-a is the room temperature sensor and the sensor 1-b is a temperature data logger installed to calculate measurement errors.

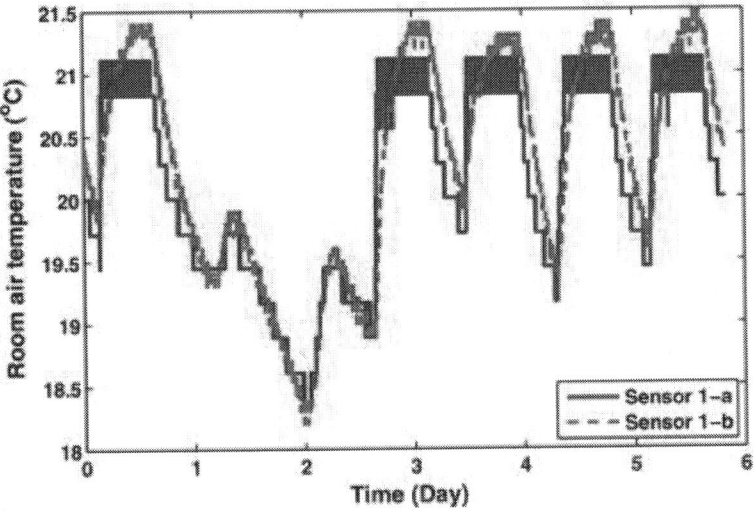

Figure. 2. Data logger and BMS sensor temperature readings in Fig. 1.

The HVAC system in the building uses ground-source heat-pumps (GSHP) to obtain required energy for heating purposes. Each unit in this system provides heating for an individual zone. Therefore, a unit operates when heating is required for its zone: the setpoint can be defined independently based on the functionality of each zone. The HVAC system uses an on–off controller to provide a desired temperature for each zone. Zone temperatures are measured with a sampling period of 60 s.

3. PARAMETER ADAPTIVE BUILDING (PAB) MODEL

Building models proposed in the literature depend on many parameters. The reason is that buildings are composed of many sub-systems and a variety of thermal mechanisms takes place in the building such as heat conduction through walls, forced convection due to air conditioning systems, and thermal radiation from outside. A mathematical model that is descriptive enough to accurately explain these phenomena will end up with many time-varying parameters. Finding the best parameters at each time step is shown to be cumbersome [23]. In this section *we propose and develop a novel parameter adaptive building (PAB) model that facilitates this parameter tuning process in an online and automatic fashion.* The architecture of the proposed PAB model is shown in Fig. 3. Measurement data from various sensors such as temperature and airflow are stored in a data repository. The PAB model has a parameter update module which takes care of automatic parameter tuning on the fly, and is explained in detail later in this section. The PAB model works as follows: historical data is used to perform off-line, one-step model calibration. The obtained parameters from model calibration is used in the parameter update module (exploiting Kalman filtering algorithm) as an initial set of parameters. Kalman filter updates the parameters of the building model, as the new measurements arrive. The control module then uses the new updated set of parameters for the next time step.

Here we first review fundamental heat transfer mechanisms in buildings, leading to a mathematical model of building climate, on top of which we develop the PAB model in the rest of this section.

3.1. Mathematical modeling

Fig. 4 depicts the schematic of a typical room studied in this paper. We use lumped model analysis [34] to reduce the complexity of the model, and obtain a low order model, suitable for control purposes. As a simplifying assumption, temperature is considered uniform inside the room. We use RC model from [22] in which the building is considered as a network of nodes. We account for *time varying* parameters by updating the parameters on the fly. More details on online parameter estimation is presented in Section 3.2

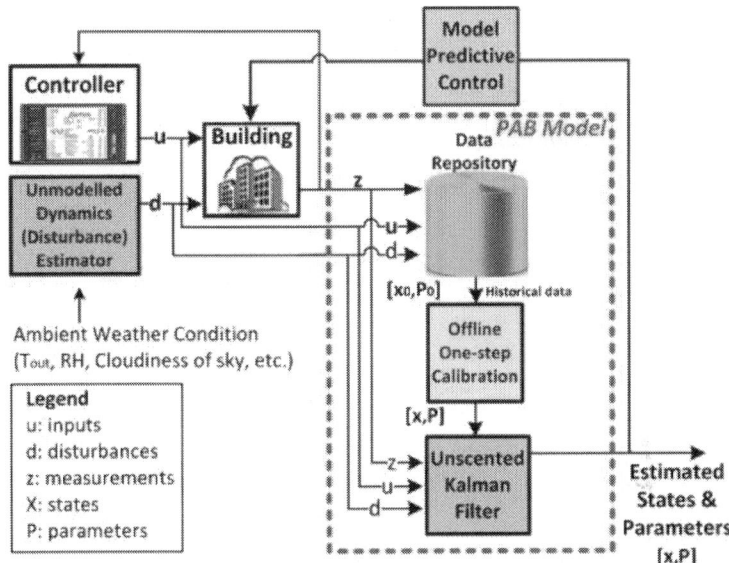

Figure. 3. Architecture of the building control based on the proposed PAB model with its components.

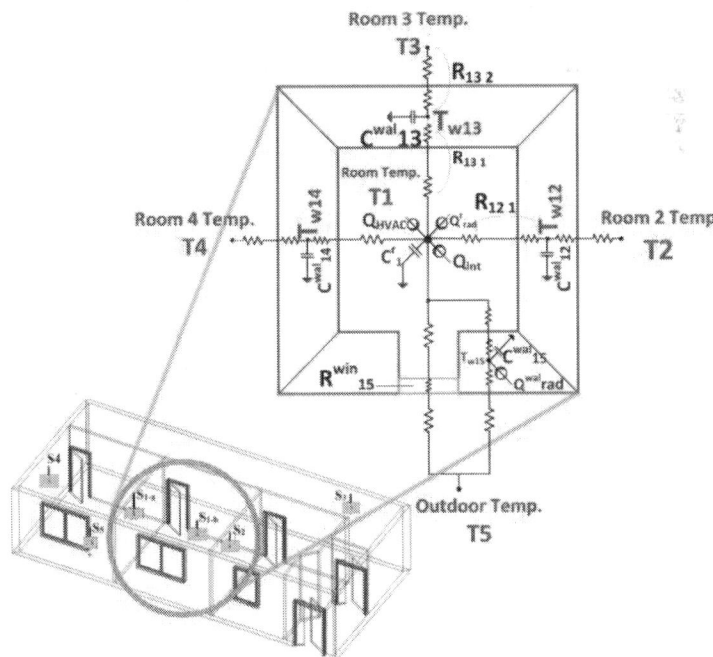

Figure 4. Schematic of a typical room with a window. Temperature sensors are denoted by "S" in this figure.

3.1.1. Heat transfer

There are two types of nodes in the building network: walls and rooms. Consider in total m nodes, m of which represent rooms and the remaining $n - m$ nodes represent walls. We denote the temperature of room i with T_{ri}. The wall node and temperature of the wall between room i and j are denoted by (i, j) and $T_{wi,j}$, respectively, thermal dynamics of which is governed by the following equation:
Equation

$$C_{i,j}^w \frac{dT_{w_{i,j}}}{dt} = \sum_{k \in \mathcal{N}_{w_{i,j}}} \frac{T_{r_k} - T_{w_{i,j}}}{R_{i,j_k}} + r_{i,j}\alpha_{i,j}A_{w_{i,j}}Q_{\mathrm{rad}_{i,j}}$$

$$(1)$$

where $C_{i,j}^w$, α_{ij} and $A_{wi,j}$ are heat capacity, radiative heat absorption coefficient and area of wall between room i and j, respectively. $R_{i,jk}$ is the total thermal resistance between the centerline of wall (i, j) and the side of the wall where node k is located. $Q_{radi,j}$ is the radiative heat flux density on wall (i, j). $N_{wi,j}$ is the set of all of neighboring nodes to node $w_{i,j}$. rij is wall identifier which is equal to 0 for internal walls, and equal to 1 for peripheral walls (i.e. either i or j is an outside node). In Eq. (1) the left term denotes the rate of change of stored heat in the wall between room i and room j. The first term of the right hand side of this equation represents the flow of heat between room k and wall (i, j) due to temperature difference and the second term shows the heat flow to the wall, due to solar radiation. Temperature dynamics of the ith room is modeled by the following equation:

$$C_i^r \frac{dT_{r_i}}{dt} = \sum_{k \in \mathcal{N}_{r_i}} \frac{T_k - T_{r_i}}{R_{i,k_i}} + \dot{m}_{r_i}c_a(T_{s_i} - T_{r_i}) + w_i\tau_{w_i}A_{\mathrm{win}_i}Q_{\mathrm{rad}_i} + \dot{Q}_{\mathrm{int}_i}$$

$$(2)$$

where T_{ri}, C_i^r and $\dot{m}_{r_i}$ are the temperature, heat capacity and air mass flow into room i, respectively. ca is the specific heat capacity of air, and T_{si} is the temperature of the supply air to room i. πi is window identifier which is equal to 0 if none of the walls surrounding room i have a window, and is equal to 1 if at least one of them has a window. τ_{wi} is the transmissivity of glass of window i, A_{wini} is the total area of windows on walls surrounding room i, Q_{radi} is the radiative heat flux density per unit area radiated to room i, and $\dot{Q}_{int_i}$ is the internal heat generation in room i. N_{ri} is the set of all of the neighboring *room* nodes to room i. In Eq. (2) the left term denotes the rate of change of stored heat in the air in room i. The first term of the right hand side of this equation represents the flow of heat between node k and room i due to temperature difference, the second term shows the heat flow delivered by the heating system, the third term represents the total radiative heat passing through the windows and the fourth term is the internal heat generation inside room i. More details of building thermal modeling and estimation of the unmodeled dynamics is available in[9], [22] and [3]. Note that we approximate the values of $Q_{radi}(t)$ and $\dot{Q}_{int}(t)$ based on the following equations:

$$Q_{radi}(t) = \tau T_{out}(t) + \zeta \tag{3}$$

$$\dot{Q}_{int}(t) = \mu \Psi(t) + \nu \tag{4}$$

where T_{out} and Ψ are the outside air temperature and CO_2 concentration in the room, respectively [35]. Air ventilation is considered constant as a simplifying assumption. A more sophisticated model for gas transport process in buildings can be found in [36]. Parameters τ, ζ, μ and ν are obtained by the parameter estimation algorithm detailed inSection 3.3.

We model the radiative heat transfer between building and ambient environment as proposed in [37]. The amount of heat transferred from the building to the environment is given by the Stefan–Boltzmann law:

$$Q_{bldg} = \epsilon \sigma T_{bldg}^4 \tag{5}$$

where T_{bldg} is the average temperature of the building. We also consider solar radiation heat transfer, Q_{solar} absorbed by the walls, and the room through the windows. The data used in this paper is based on the past 30 years monthly average of solar radiation for flat-plate collectors facing south (resembling the south facing flat vertical walls of the building), and is obtained from NREL (National Renewable Energy Laboratory) [38] database for Houghton, MI in January. Furthermore, we take into account the radiation cooling at night (i.e. sky thermal radiation to the building) based on the proposed relation in [37]:

$$Q_{sky} = (1 + KC^2)8.78 \times 10^{-13} T_{out}^{5.852} RH^{0.07195} \tag{6}$$

where K is the coefficient related to the cloud height and C is a function of cloud coverage. We use $K = 0.34$ and $C = 0.8$ for simulations, based on the results in [37]. T_{out} is the outside air temperature, and RH is the air relative humidity percentage.

The total radiation exchange between building and ambient environment is then given by:

$$Q_{rad} = Q_{sky} + Q_{solar} - Q_{bldg} \tag{7}$$

Note that Q_{sky} and Q_{solar} are heat flow *into* the building, and Q_{bldg}, is the heat flow *from* the building *to* the environment.

3.1.2. System dynamics

Heat transfer equations for walls and rooms yield the following system dynamics:

$$\dot{x}_t = f(x_t, u_t, d_t, t)$$
$$y_t = Cx_t \tag{8}$$

where $x_t \in \mathbb{R}^n$ is the state vector representing the temperature of the nodes in the thermal network, $u_t \in \mathbb{R}^{lm}$ is the input vector representing the air mass flow rate and discharge air temperature of conditioned air into each thermal zone, and $y_t \in \mathbb{R}^m$ is the output vector of the system which represents the temperature of the thermal zones. l is the number of inputs to each thermal zone (e.g., two for air mass flow and supply air temperature). C is a matrix of proper dimension and the disturbance vector is given by $d_t = g(Q_{rad_i}(t), \dot{Q}_{int_i}(t), T_{out}(t))$.

3.1.3. Disturbance

Following the intuitive linear relation between outside temperature T_{out}, internal heat generation $\dot{Q}_{int}$, and solar radiation Q_{rad}, with the building internal temperature rise we approximate d_t with an affine function of these quantities, leading to:

$$d_t = aQ_{rad_i}(t) + b\dot{Q}_{int}(t) + cT_{out}(t) + e \tag{9}$$

where a, b, c, e are constants to be estimated. By substituting (3) and (4) into (9) and rearranging the terms, we get:

$$d_t = (a\tau + c)T_{out}(t) + b\mu\Psi(t) + a\zeta + bv + e = \bar{a}T_{out}(t) + \bar{b}\Psi(t) + \bar{e} \tag{10}$$

where $\bar{a} = a\tau + c, \bar{b} = b\mu$, and $\bar{e} = a\zeta + bv + e$. Therefore, only measurements of outside air temperature and CO_2 concentration levels are needed to determine the disturbance to the model. The values of $\bar{a}, \bar{b}$, and $\bar{e}$ are estimated along with other parameters of the model.

3.1.4. Additive uncertainty

We linearize the original nonlinear dynamic system and use Euler's discretization method to obtain a linear discrete-time system. We also add an additive uncertainty to the state update equation to account for model uncertainties, leading to:

$$x_{k+1} = Ax_k + Bu_k + E(d_k + w_k) \tag{11}$$

where the uncertainty $w_k \in \mathbb{R}^r$ is a stochastic additive disturbance. $k \in \mathbb{Z}$ refers to time in continuous-time domain and $k \in \mathbb{Z}$ refers to time in discrete-time domain. The set of possible disturbance uncertainties is denoted by W and $w_k \in \mathcal{W}$ $\forall k = 0, 1, \ldots, N-1.$. For this study, we consider box-constrained disturbance uncertainties given by

$$\mathcal{W}_\lambda = \{w : \|w\|_\infty \le \lambda\} \tag{12}$$

3.2. State-parameter estimation

Using (1) for each wall and (2) for each room node in the building network, system dynamics is given by:

$$\dot{x}_1 = \frac{1}{C_1} \cdot \left(\left(\frac{1}{R_{12_1}} - \frac{1}{R_{13_1}} - \frac{1}{R_{14_1}} - \frac{1}{R_{15_1}} - \frac{1}{R_{15}^{win}} - \dot{m}_{r1}c_a \right) x_1 \right.$$
$$\left. + \frac{x_2}{R_{12_1}} + \frac{x_3}{R_{13_1}} + \frac{x_4}{R_{14_1}} + \frac{x_5}{R_{15_1}} + c_a T_{s1}\dot{m}_{r1} + \frac{T_5}{R_{15}^{win}} + A_{win}\tau Q_{rad} + \dot{Q}_{int_1} \right) \tag{13a}$$

$$\dot{x}_2 = \frac{1}{C_{21}^w} \cdot \left(\frac{x_1}{R_{21_1}} - \left(\frac{1}{R_{21_1}} + \frac{1}{R_{21_2}} \right) x_2 + \frac{T_2}{R_{21_2}} \right) \tag{13b}$$

$$\dot{x}_3 = \frac{1}{C_{31}^w} \cdot \left(\frac{x_1}{R_{31_1}} - \left(\frac{1}{R_{31_1}} + \frac{1}{R_{31_3}} \right) x_3 + \frac{T_3}{R_{31_3}} \right) \tag{13c}$$

$$\dot{x}_4 = \frac{1}{C_{41}^w} \cdot \left(\frac{x_1}{R_{41_1}} - \left(\frac{1}{R_{41_1}} + \frac{1}{R_{41_4}} \right) x_4 + \frac{T_4}{R_{41_4}} \right) \tag{13d}$$

$$\dot{x}_5 = \frac{1}{C_{51}^w} \cdot \left(\frac{x_1}{R_{51_1}} - \left(\frac{1}{R_{51_1}} + \frac{1}{R_{51_5}} \right) x_5 + \frac{T_5}{R_{51_5}} + A_{w_{51}}\alpha Q_{rad} \right) \tag{13e}$$

where x_1 is the room temperature (T_{r1}), and x_2, x_3, x_4, x_5 are the peripheral walls' temperature (i.e. $T_{w12}, T_{w13}, T_{w14}, T_{w15}$). T_2, T_3, T_4, T_5 are the temperatures of the surrounding zones, as shown in Fig. 4 and Fig. 5. These temperatures act as disturbance to the system dynamics for a single zone thermal model, and x is the state vector:

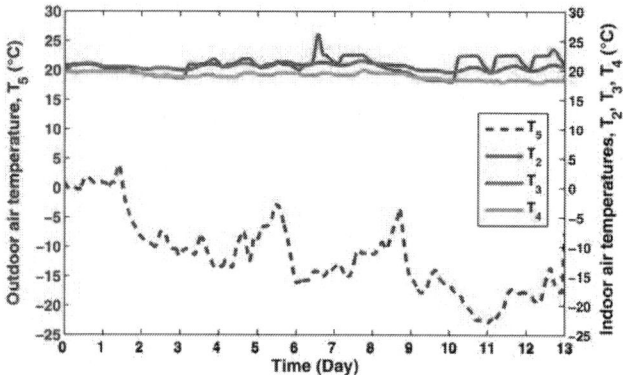

Figure 5. Temperatures of neighboring zones acting as disturbance to the PAB model.

$$x = \begin{bmatrix} T_{r1}, T_{w12}, T_{w13}, T_{w14}, T_{w15} \end{bmatrix}^T \tag{14}$$

One way to adapt the model to account for time varying parameters is to assume that all the parameters of the model are independent, and hence define a state corresponding to each parameter. However, this would lead to excessive number of states (e.g. 18 states for a room shown in Fig. 4). To overcome this problem, we take a different approach. We reduce the number of states by exploiting the redundancies in the resulting model. For instance, thermal properties of wall material (e.g. specific heat capacity and conductive heat transfer coefficient) are the same across the building, as these are functions of the materials used as the building walls. In addition, the thickness of internal walls and thickness of peripheral walls are the same throughout the building. Following this approach, we are able to reduce the number of independent parameters from 18 to 10. Hence we re-write the thermal equations of the walls, i.e. (13b), (13c), (13d) and (13e) as follows:

$$\dot{x}_2 = \frac{x_1}{C^w R^w} - \frac{2}{C^w R^w} x_2 + \frac{T_2}{C^w R^w} \tag{15}$$

$$\dot{x}_3 = \frac{x_1}{C^w R^w} - \frac{2}{C^w R^w} x_3 + \frac{T_3}{C^w R^w} \tag{16}$$

$$\dot{x}_4 = \frac{x_1}{C^w R^w} - \frac{2}{C^w R^w} x_4 + \frac{T_4}{C^w R^w} \tag{17}$$

$$\dot{x}_5 = \frac{x_1}{C_{51}^w R_{51_1}} - \left(\frac{1}{C_{51}^w R_{51_1}} + \frac{1}{C_{51}^w R_{51_5}} \right) x_5 + \frac{T_5}{C_{51}^w R_{51_5}} + \frac{A_{w51} \alpha Q_{rad}}{C_{51}^w} \quad (18)$$

As shown in (19), $C_w R_w$ is not a function of the area of wall:

$$C^w R_w = (c_w A_w L_w) \left(\frac{L_w/2}{k_w A_w} + \frac{1}{h_{in} A_w} \right) = \frac{c_w L_w^2}{2k_w} + \frac{c_w L_w}{h_{in}} \quad (19)$$

where c_w, k_w, A_w and L_w are the specific heat capacity, conductive heat transfer coefficient of wall material, area and thickness of wall, respectively, and h_{in} is the indoor convective heat transfer coefficient. Hence, we can use one common term to express thermal capacitance–resistance between centerline of each wall and the node on each side of the wall for the equations of walls in the building.

We designate a state variable to all the independent time-varying parameters of the system as follows:

$$x_6 = \frac{1}{C_1^r R_{12_1}} \quad x_7 = \frac{1}{C_1^r R_{13_1}} \quad (20)$$

$$x_8 = \frac{1}{C_1^r R_{14_1}} \quad x_9 = \frac{1}{C_1^r R_{15_1}} \quad (21)$$

$$x_{10} = \frac{1}{C_1^r} \quad x_{11} = \frac{1}{C_w R_w} \quad (22)$$

$$x_{12} = \frac{1}{C_{51}^w R_{51_1}} \quad x_{13} = \frac{1}{C_{51}^w R_{51_5}} \quad (23)$$

$$x_{14} = \frac{\alpha}{C_{51}^w} \quad x_{15} = \frac{1}{R_{15}^{win}} \quad (24)$$

Rate of change of these states is equal to zero, as shown in the corresponding state update Eq. (30). We then add a low-magnitude *fictitious noise* to the dynamics of parameters to allow slow changes in their values over time.

$$\dot{x}_1 = (x_6 - x_7 - x_8 - x_9 - x_{10}x_{15} - x_{10}u_2 C_a)x_1 + x_6 x_2 + x_7 x_3$$
$$+x_8 x_4 + x_9 x_5 + (C_a u_1 u_2 + T_5 x_{15} + A_{win}\tau Q_{rad} + \dot{Q}_{int})\cdot x_{10} \qquad (25)$$

$$\dot{x}_2 = (x_1 - 2x_2 + T_2)\cdot x_{11} \qquad (26)$$

$$\dot{x}_3 = (x_1 - 2x_3 + T_3)\cdot x_{11} \qquad (27)$$

$$\dot{x}_4 = (x_1 - 2x_4 + T_4)\cdot x_{11} \qquad (28)$$

$$\dot{x}_5 = x_1 x_{12} - (x_{12} + x_{13})x_5 + T_5 x_{13} + A^{W_{51}} x_{14} Q_{rad} \qquad (29)$$

$$\dot{x}_i = 0 \quad \forall i = 6, 7, \ldots 15. \qquad (30)$$

u is the input vector given by:

$$u = \begin{bmatrix} T_{s1} \\ \dot{m}_{r1} \end{bmatrix} \qquad (31)$$

In summary, we express the dynamics of the system using following state update model:

$$\begin{aligned} x_k &= f(x_{k-1}, u_{k-1}, d_{k-1}, w_{k-1}) \\ z_k &= h(x_k) + v_k \end{aligned} \qquad (32)$$

where w_k and v_k are the process and measurement noise and are assumed to be zero mean multivariate Gaussian process with variance Wk and Vk, (i.e. $w_k \sim N(0, W_k)$ and $v_k \sim N(0, V_k)$), respectively.

3.3. Estimation algorithm

In order to estimate the unknown parameters of the system we augment the states of the system with a vector pk which stores the parameters of the system, with a time evolution dynamics of $pk_{+1} = pk$, as will be detailed in Appendix B. Due to the multiplication of states and parameters the resulting dynamic system is nonlinear. Nonlinear estimation algorithms such as extended Kalman filtering (EKF) or unscented Kalman filtering (UKF) can then be exploited to simultaneously estimate the states and the parameters of the system.

An alternative to using a Kalman filter would be a simple observer. However, given the random variations, inaccuracies and uncertainties in the system dynamics, as described earlier in the paper, using a Kalman filter is suggested in order to get a statistically optimal estimate of system states [39] and [40].

In our previous work [41] we showed that UKF outperforms EKF for building parameter estimation. Thus, we only focus on UKF in this study. We present an algorithmic description of the UKF in Appendix B, omitting some theoretical considerations.

3.3.1. Estimation results

The test-bed from Section 2 was used to collect measurements from January 11 to January 24, 2013. To remove noise from the temperature measurements, a second order Butterworth lowpass filter with cutoff frequency of 0.001 Hz was used. Fig. 5 shows the temperatures of the neighboring zones and the outside temperature which act as disturbance to the PAB model. Fig. 6 depicts the model inputs including the air mass flow rate and the supply air temperature. In order to obtain the best initial parameter values for the Kalman filter algorithm, we first perform a (static) parameter identification on the historical data. We consider the first part of the data as *training set* (shown in red in Fig. 7), and obtain the best parameters that minimize the least square error between the simulation and the measurement data. The result of this step is used to simulate the temperature evolution of the room air for the next three days (shown in black in Fig. 7). Due to time-varying parameters and disturbance to the model, it is difficult to find a set of parameters for the model which results in good temperature tracking for all days including weekdays and weekends, and hence, as shown in Fig. 7, the results of simulations for the following days in the *testing data* set is even worse.

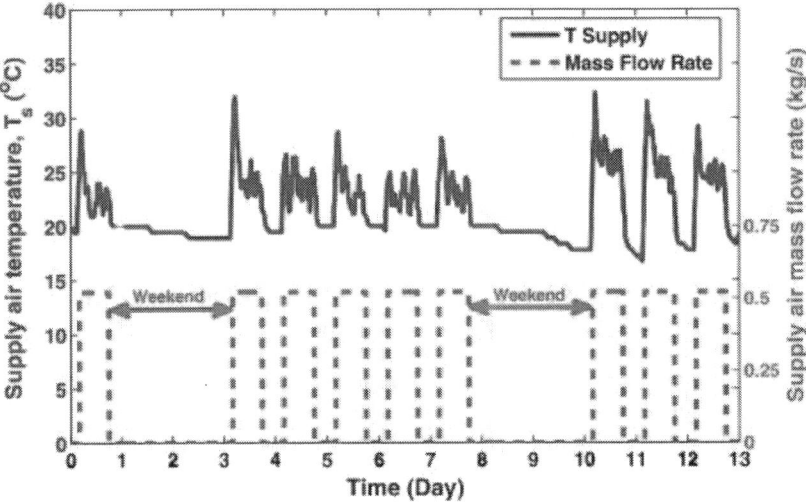

Figure 6. Inputs to the PAB model.

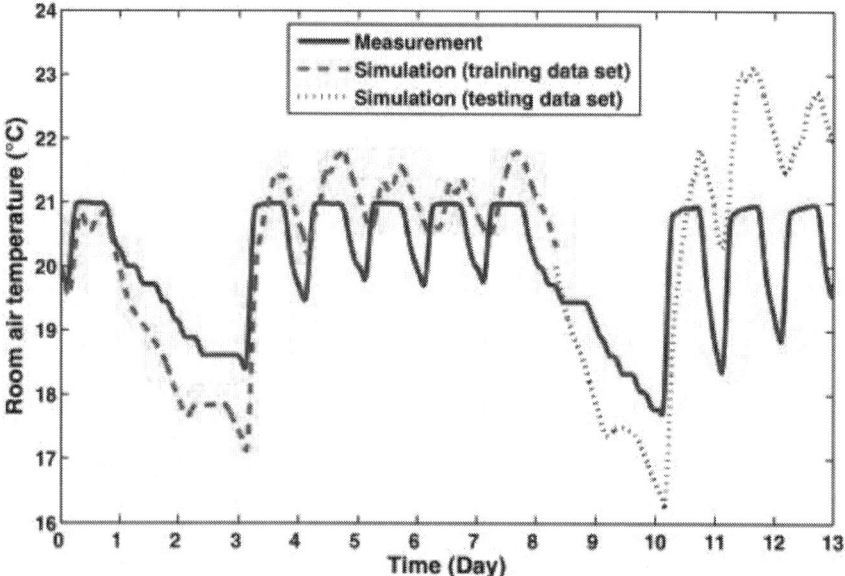

Figure 7. Off-line parameter calibration of the PAB model using room temperature measurements. The first set of data (shown in red) is the training data. We identify the parameters in one shot optimization by minimizing the ℓ_2 norm of the error between simulation and measurement data. Then we used the obtained parameters from the training data set (off-line calibration results) to predict the temperature evolution for the next days (shown in black). (For interpretation of the references to color in this figure legend, the reader is referred to the web version of the article.)

The obtained initial parameters from the off-line calibration step is used as initial value for the UKF algorithm. For the off-line parameter calibration practice, we used the historical data of two weeks where the first 60% of the data was used for training (calibration) and the remaining 40% of data was used for testing. The temperature estimation of room and walls, using UKF are depicted in Fig. 8 and Fig. 9. The evolution of parameters over time is shown in Fig. 10. The parameters evolve over time and the steady state values are not necessarily close to the initial points as expected, due to the changing environment. Note that the first part of the estimation of wall temperature by UKF leads to overshoot in the wall temperature, however, this overshoot is quickly recovered as UKF uses more data to tune the parameters more accurately.

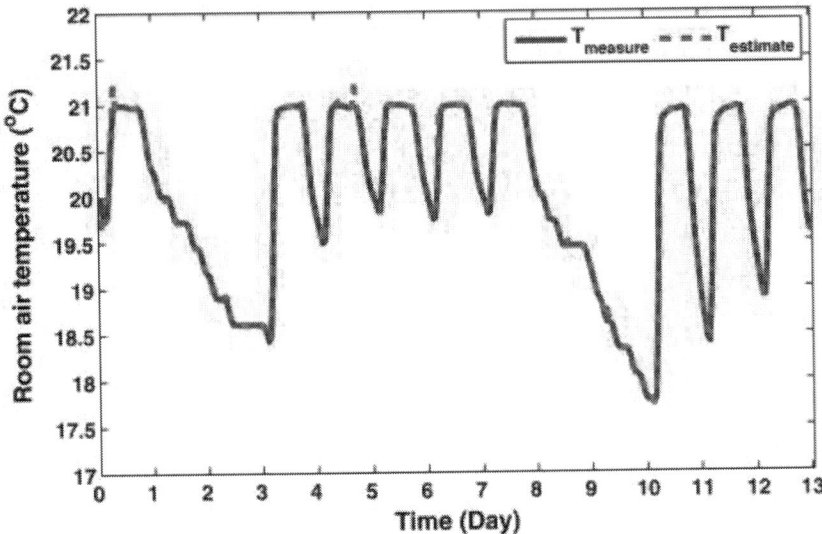

Figure 8. Estimated and measured room temperature using the designed UKF.

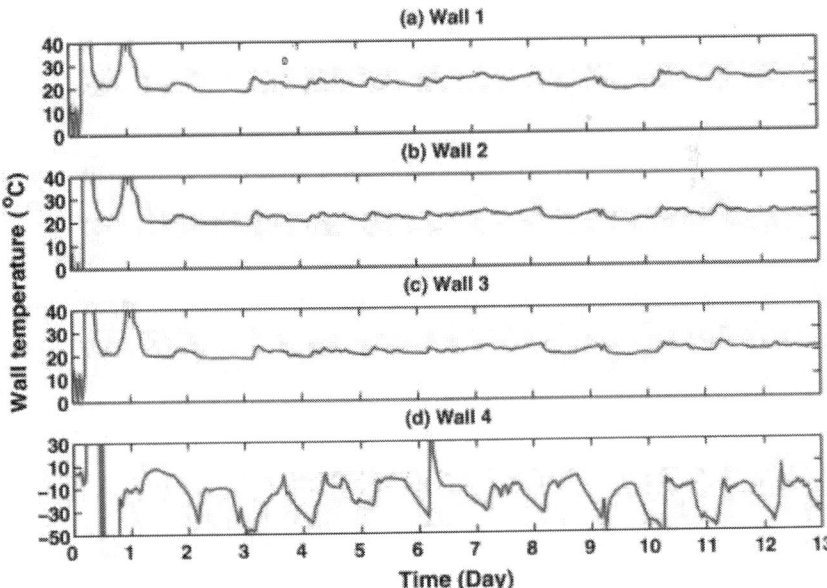

Figure 9. Estimated temperature of walls using UKF. We have zoomed the figures to focus on the more steady estimates of the walls rather than the first part transient behavior.

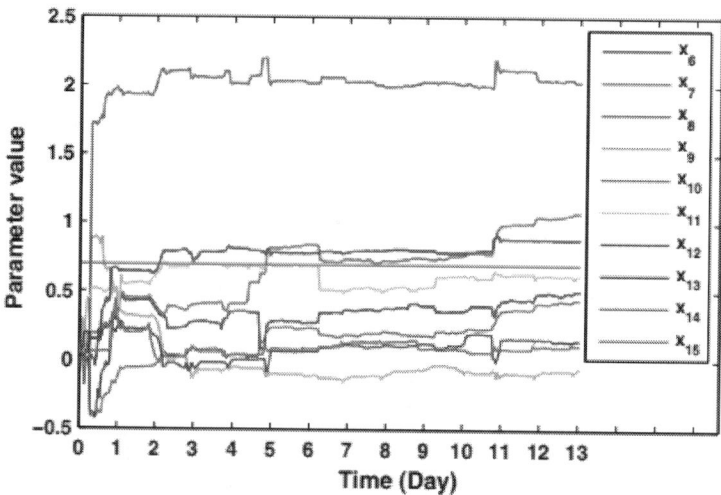

Figure 10. Estimated parameters of the PAB model using the designed UKF.

UKF is also tested to estimate the temperature in the presence of process and measurement noise (w and v, respectively) as shown in Fig. 11. We add process and measurement noise to the model and use UKF to estimate the temperatures. UKF is used to estimate the temperature from the measurements. Performance of UKF is shown with model uncertainty w, and measurement noise v, given by $w_k \sim N(0,0.2)$ and $v_k \sim N(0,1.4)$, respectively. As seen in Fig. 11, UKF is able to cancel out the effect of noise very effectively.

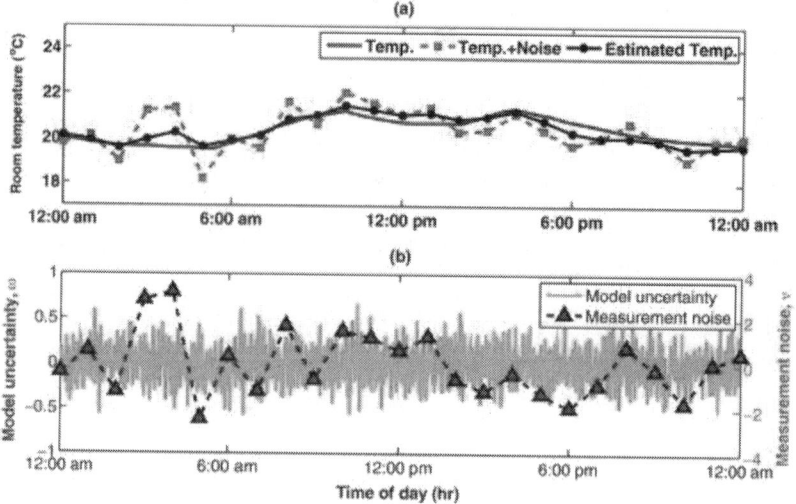

Figure 11. Performance of the designed UKF in the presence of model uncertainty and measurement noise.

4. CONTROLLER DESIGN

In this section we study *the impact of the use of the PAB model in a model-based control design framework*. State-of-the-art is to use a fixed-parameter model to design MPC for buildings. We propose using the updated parameter model obtained using the Kalman filter estimation process at each time step as shown in Fig. 3, which results in a more accurate model and hence lower model uncertainty. The underlying assumption here is that the parameters of the system do not change from time t to $t + 1$. At the next time step, MPC uses the model with updated parameters, to derive the optimal inputs. Inputs are implemented on the system and at the next sampling time new states (temperatures) are measured and sent to the PAB model, and this process repeats.

We also formulate a nominal MPC and a robust model predictive control (RMPC), and study their performances for various model uncertainty levels. MPC assumes that the model is perfect (no uncertainty), and the RMPC assumes that the model is uncertain and designs a robust control policy for a specific class of uncertainty. The results from MPC and RMPC are compared to a conventional rule-based control (RBC) for a typical building. *Novel performance indices are proposed* to compare the performance of these controllers. We also present *a methodology to select the best controller* among the ones studied in this section for any given model uncertainty, which leads to optimum trade-off between energy consumption and comfort level.

4.1. ASHRAE requirements for building climate control

ASHRAE's Standard 55 [42], *Thermal Environmental Conditions for Human Occupancy*, suggests the condition which is acceptable to at least 80% of occupants. According to this standard, the ideal temperature in typical clothing in summer (0.35–0.6 clo) is in the range of 22.5–26 °C. The operative temperature for occupants in normal clothing insulation in winter which is between 0.8 and 1.2 clo should be in range of 20–23.5 °C. This temperature range is based on a metabolic rate of 1.2 met (70 W/m²) and 60% RH. More details can be found in [43], [44] and [45]. ASHRAE's Standard 62.1 [46], *Ventilation for Acceptable Indoor Air Quality*, explains outdoor air ventilation requirements for different types of indoor spaces. When the major contamination source is proportional to number of occupants, the *minimum ventilation rate* is enforced in CFM (L/s) and when other factors play the main role in contamination, the *minimum ventilation rate* is enforced in CFM/ft² (L/s m²) [45]. We use this as a guideline for control design in this section.

4.2. Rule-based control (RBC)

The rule-based controller in this study is a conventional on–off HVAC controller. The time constant of the control action implementation is $\Delta t = 1$ h. The controller opens the dampers of conditioned air flow to the thermal zones when heating is required and keeps it fully open for the duration of Δt. In the next time step the controller checks the temperature again and adjusts the damper position if the room temperature is within the comfort zone, or keeps it open if the room air temperature is still outside the comfort zone. In on–off control, position of the dampers can be either the *min* value or the *max* value. When system goes to the cooling mode, supply air temperature changes accordingly. The experimental data presented here is for the heating mode only. To be consistent and to perform a fair comparison, we use the same time constants Δt for all controllers.

4.3. Model predictive control (MPC)

A model predictive control problem is formulated with the objective of minimizing a linear combination of the total and the peak airflow. We implement the control inputs obtained from the MPC with the linearized system dynamics of the model on the original nonlinear model for forward simulation.

Fan energy consumption is proportional to the *cubic* of the airflow. Hence minimizing the peak airflow would dramatically reduce fan energy consumption. We have considered a cost function for the MPC which comprises linear combination of the total input airflow (ℓ_1 norm of input) and the peak of airflow (ℓ_∞ norm of input). The alternative would be to use the actual nonlinear function of fan energy consumption. However, it would lead to nonlinear MPC which is much slower than linear MPC. We use the proposed cost function to achieve better computational properties. Also in order to guarantee feasibility (constraint satisfaction) at all times, we implement soft constraints. The predictive controller solves at each time step the following optimization problem:

$$\min_{U_t, \bar{\epsilon}, \underline{\epsilon}} \{\|U_t\|_1 + \kappa|U_t|_\infty + \rho(|\bar{\epsilon}_t|_1 + |\underline{\epsilon}_t|_1)\} = \tag{33a}$$

$$\min_{U_t, \bar{\epsilon}, \underline{\epsilon}} \{\sum_{k=0}^{N-1} |u_{t+k|t}| + \kappa \max(|u_{t|t}|, \ldots m, |u_{t+N-1|t}|)$$
$$+ \rho \sum_{k=1}^{N} (|\bar{\epsilon}_{t+k|t}| + |\underline{\epsilon}_{t+k|t}|)\} \tag{33b}$$

$$\text{s.t. } x_{t+k+1|t} = Ax_{t+k|t} + Bu_{t+k|t} + Ed_{t+k|t}, \quad k = 0, \ldots, N-1 \tag{33c}$$

$$y_{t+k|t} = Cx_{t+k|t}, \quad k = 1, \ldots, N \tag{33d}$$

$$\underline{u}_{t+k|t} \leq u_{t+k|t} \leq \overline{u}_{t+k|t}, \quad k = 0, \ldots, N-1 \tag{33e}$$

$$\underline{T}_{t+k|t} - \underline{\varepsilon}_{t+k|t} \leq y_{t+k|t} \leq \overline{T}_{t+k|t} + \overline{\varepsilon}_{t+k|t}, \quad k = 1, \ldots, N \tag{33f}$$

$$\underline{\varepsilon}_{t+k|t}, \overline{\varepsilon}_{t+k|t} \geq 0, \quad k = 1, \ldots, N \tag{33g}$$

where $U_t = [u_{t|t}, u_{t+1|t}, \ldots, m, u_{t+N-1|t}]$ is vector of control inputs, and $\underline{\epsilon} = [\underline{\varepsilon}_{t+1|t}, \ldots m, \underline{\varepsilon}_{t+N|t}]$ and $\overline{\epsilon} = [\overline{\varepsilon}_{t+1|t}, \ldots m, \overline{\varepsilon}_{t+N|t}]$ are the slack variables used to utilize soft constraints on room temperature. $y_{t+k|t}$ is the room temperature vector, $d_{t+k|t}$ is the disturbance load prediction, and $\underline{T}_{t+k|t}$ and $\overline{T}_{t+k|t}$ for $k = 1, \ldots m, N$ are the lower and upper limits on the room temperature, respectively. $\underline{u}_{t+k|t}$ and $\overline{u}_{t+k|t}$ are the lower and upper limits on the airflow input by the variable air volume (VAV) damper, respectively. Note that based on the ASHRAE Standard 62.1 – Section 6.2.6.1, during unoccupied hours, ventilation systems should be able to maintain the required non-zero ventilation rates $(\underline{u}_{t+k|t} > 0)$ in the breathing zone [46]. ρ is the penalty on the comfort constraint violations, and κ is the penalty on peak power consumption.

At each time step only the first entry of U_t is implemented on the model. At the next time step the prediction horizon N is shifted leading to a new optimization problem. The prediction horizon is $N = 24$, and at each time step only the first entry of the input vector U_t is implemented on the model. This process is repeated over and over until the total time span of interest is covered. We use YALMIP [47] to set up the MPC problem in MATLAB.

4.4. Robust model predictive control (RMPC)

We consider additive uncertainty to the system model as previously described in (11). A schematic of the robust optimal control implementation on the nonlinear building model is shown in Fig. 12. In RMPC algorithm, the cost function is the same as in the one in MPC case:

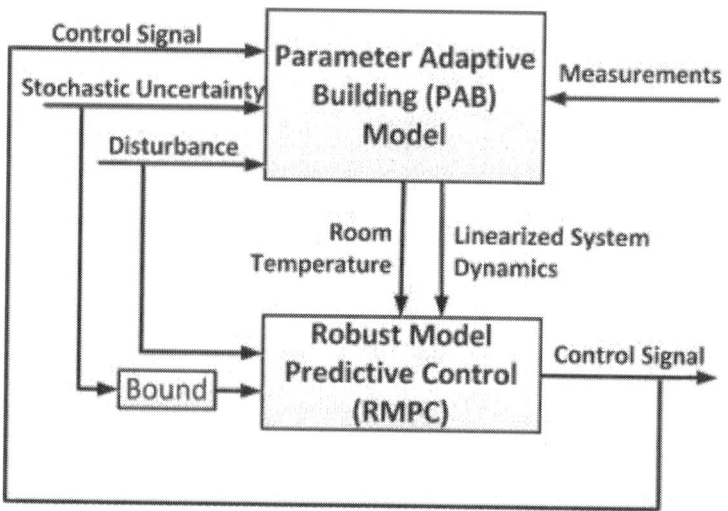

Figure 12. Schematic of the robust model predictive control implementation.

$$\min_{U_t,\bar{\varepsilon},\underline{\varepsilon}} \{||U_t||_1 + \kappa||U_t||_\infty + \rho(||\bar{\epsilon}_t||_1 + ||\underline{\epsilon}_t||_1)\}$$

$$(34)$$

However, state and input constraints are as follows:

$$x_{t+k+1|t} = Ax_{t+k|t} + Bu_{t+k|t} + E(d_{t+k|t} + w_{t+k|t}) \quad k = 0, 1, \ldots, N-1 \quad (35a)$$

$$y_{t+k|t} = Cx_{t+k|t} \quad k = 1, 2, \ldots, N \tag{35b}$$

$$\underline{T}_{t+k|t} - \underline{\varepsilon}_{t+k|t} \le y_{t+k|t} \le \bar{T}_{t+k|t} + \bar{\varepsilon}_{t+k|t} \quad k = 1, 2, \ldots, N \tag{35c}$$

$$\underline{u}_{t+k|t} \le u_{t+k|t} \le \bar{u}_{t+k|t} \quad k = 0, 1, \ldots, N-1 \tag{35d}$$

$$\underline{\varepsilon}_{t+k|t}, \quad \bar{\varepsilon}_{t+k|t} \ge 0 \quad k = 1, 2, \ldots, N \tag{35e}$$

$$\forall w_{t+k|t} \in W \quad k = 0, 1, \ldots, N-1 \tag{35f}$$

The only difference with respect to MPC algorithm is the introduction of additive uncertainty term w in the state update equation.

Using this formulation, we derive a robust counterpart of an uncertain optimization problem in which constraints are satisfied for all possible uncertainties, and worst-case objective is calculated.

It is shown in [16] that the open-loop constrained robust optimal control (OL-CROC) is conservative. The closed-loop constrained robust optimal control

(CL-CROC) formulation overcomes this issue but it can quickly lead to an intractable problem [48]. Next, we review the feedback prediction concept followed by our proposed formulation to improve upon the feedback prediction scheme.

4.4.1. Feedback predictions

The idea in feedback prediction, is to introduce new decision variables and parameterize the future control sequences using the future disturbances and an additive independent decision variable.

Define an affine disturbance feedback as:

$$u_i := \sum_{j=0}^{i-1} m_{i,j} w_j + n_i \quad \forall i = 1, 2, \ldots, N - 1 \tag{36}$$

Therefore the input vector can be written as $U = \mathbf{M}\mathbf{w} + \mathbf{n}$, where $\mathbf{M}$ and $\mathbf{n}$ are given by

$$\mathbf{M} := \begin{bmatrix} 0 & \ldots m & \ldots m & 0 \\ m_{1,0} & 0 & \ddots & 0 \\ \vdots & \ddots & \ddots & \vdots \\ m_{N-1,0} & \ldots m & m_{N-1,N-2} & 0 \end{bmatrix}, \quad \mathbf{n} := \begin{bmatrix} n_0 \\ \vdots \\ \vdots \\ n_{N-1} \end{bmatrix} \tag{37}$$

and the vector of disturbances is given by $\mathbf{w} = \begin{bmatrix} w_0 & w_1 & \cdots & w_{N-1} \end{bmatrix}'$.

The control sequence is parameterized directly in the uncertainty. What we have here is basically a sub-optimal version of the closed-loop min-max solution [48].

4.4.2. Two-lower-diagonal structure (TLDS)

One of the parameterizations introduced in [48] is Lower Triangular Structure (LTS). The main problem with the min–max formulation based on LTS parameterization is the excessive number of decision variables and constraints. The reason is the high-dimensional parameterization of matrix $\mathbf{M}$. To resolve the issue of high-dimensional parameterization of matrix $\mathbf{M}$, we propose the following new parameterizations.

By analyzing the structure of the optimal matrix $\mathbf{M}$, it was observed that the parameterization of the input does not need to consider feedback of more than past two values of $\mathbf{w}$ at each time, hence we propose the following disturbance feedback.

$$u_i \quad := m_{i,i-2}w_{i-2} + m_{i,i-1}w_{i-1} + n_i$$

$$= \sum_{j=i-2}^{i-1} m_{i,j}\omega_j + n_i \quad \forall i = 1, 2, \ldots, N-1$$

$$(38)$$

and the corresponding parameterization matrix $\mathbf{M}$ is an $N \times N$ matrix that has the entries on the first and second diagonal of $\mathbf{M}$ below its main diagonal as decision variables and 0 elsewhere. $\mathbf{n}$ remains as in (37). With this structure we exploit the sparsity of the feedback gain matrix to enhance the computational characteristics of the controller.

4.5. Performance indices

To compare the overall performance of the proposed controllers we define indices to measure the energy consumption and comfort level provided by each controller. In addition, we define a new index to evaluate the overall performance of each controller considering both the energy and comfort indices.

- The energy index Ie in (kWh) is defined:

$$I_e = \int_{t=0}^{24} [P_c(t) + P_h(t) + P_f(t)]\, dt$$

$$(39)$$

where cooling power Pc, heating power Ph and fan power Pf are determined by:

$$P_c(t) = \dot{m}_c(t)c_p[T_{out}(t) - T_c(t)]$$

$$(40a)$$

$$P_h(t) = \dot{m}_h(t)c_p[T_h(t) - T_{out}(t)]$$

$$(40b)$$

$$P_f(t) = \alpha_3 \dot{v}^3(t) + \alpha_2 \dot{v}^2(t) + \alpha_1 \dot{v}(t) + \alpha_0$$

$$(40c)$$

$cp = 1.012$ (kJ/kg °C)   is   the   specific   heat   capacity   of   air and $\alpha_3 = -6.06 \times 10^{-13}$ (CFM^{-3})and $\alpha_2 = 0.73 \times 10^{-8}$ (CFM^{-2})and $\alpha_1 = -1.2 \times 10^{-3}$ (CFM^{-1}) and $\alpha_0 = 59.2$ and $\dot{v} = \dot{m}/\rho$ is in (CFM), where ρ is density of air and $P_f(t)$ is in % of nominal power consumption [49]. Using these constants, the fan power values, in (kW), can be calculated.

- The discomfort index Id in degree Celsius hour (°Ch) is defined as the sum of all the temperature violations in the course of a day.

$$I_d = \int_{t=0}^{24} \left[\min \left\{ |T(t) - \overline{T}(t)|, |T(t) - \underline{T}(t)| \right\} . \mathbf{1}_{B(t)^c}(T(t)) \right] dt \tag{41}$$

where $B(t) = [\underline{T}(t), T(t)]$ is the comfort zone at time t and $\mathbf{1}$ is the indicator function.

- A good control performance means not only low energy consumption, but also low resulting discomfort. To assess the overall performance of the controllers, we need to examine both Ie and Id at the same time. Using the two indices defined above we define a third index called overall performance index (IOP). The intuition behind this new index is to take into account the energy and discomfort index in one single term. IOP is defined as:

$$I_{OP} = \frac{(I_d^* - I_d)/\|I_d\|_\infty}{I_e/\|I_e\|_\infty} \tag{42}$$

where I_d^* is the maximum allowed discomfort and $\|.\|_\infty$ denotes infinity norm or the maximum value of energy indices among all three controllers. Negative value of IOP means that the discomfort index is not within the preferred range. The lower the Id and Ie are, the higher the IOP will be. Therefore, the higher the IOP, the better the overall performance. In this study, the limit on the allowed discomfort index is heuristically chosen to be $I_d^* = 0.5$ (°Ch) to ensure adequate comfort level.

4.6. Control results

To illustrate the effectiveness of the controllers proposed in 4.3 and 4.4, we assess their performances for different *model uncertainty* values denoted by δ and defined as

$$\delta = \frac{\lambda}{\|\mathbf{d}\|_\infty} \times 100(\%) \tag{43}$$

where λ is the ℓ_∞ norm bound of the uncertainty as given by (12) and $\mathbf{d} = [d'_1, d'_2, \ldots, d'_N]'$ is the disturbance realization vector. d' represents transpose of vector d.

A time constant of $\Delta t = 1$ (h) is used for all controllers. We implement the introduced model predictive controllers with a prediction horizon of $N = 24$. The choice of $N = 24$ is to provide a good balance between performance and computational cost for the MPC framework in this study. We use the following numerical values for parameters in (4.3) and (4.4). $\overline{u} = 63$ cfm ($0.03 \text{ m}^3/\text{s}$) is the higher limit on air mass flow, $[\underline{T}_{,t} \ \overline{T}_{,t}] = [20 \ 22]$ °C during occupied hours, and $[\underline{T}_{,t} \ \overline{T}_{,t}] = [19 \ 23]$ °C is used during unoccupied hours. For the simulations we use $\kappa = 0.75$

and $\rho = 50$. $\underline{\varepsilon}$ and $\overline{\varepsilon}$ are the slack variables used to avoid feasibility problem, where $\underline{\varepsilon}$ and $\overline{\varepsilon}$ are the vectors storing slack variables.

Optimal controller and the resulting room temperature with the presence of a box-constrained uncertainty in four cases are depicted in Fig. 13. Measurements, as shown in black, shows the air mass flow and temperature recording for the room using a simple existing control policy of the building HVAC system. RBC represents the result of the rule-based control. MPC refers to the performance of a model-based control algorithm in which no knowledge of the model uncertainty is known a-prior to the control algorithm. RMPC refers to the simulation of the control algorithm which considers the model uncertainty bound and utilizes the uncertainty feedback strategy of (36) in designing the control policy.

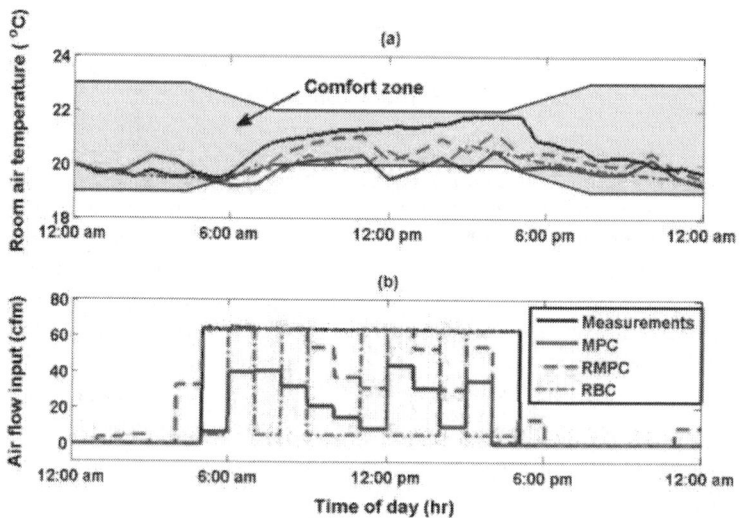

Figure 13. Control input and resulting temperature profile for the existing controller on the building (measurements), RBC, MPC, and robust MPC controllers. The additive uncertainty bound is considered $\delta = 60\%$ in this case.

We consider stochastic uncertainties with different uncertainty bounds (λ) as introduced in (12). The MPC does not have any a-priori information regarding the additive uncertainty, and calculates the controller solely based on the deterministic system dynamics. However the RMPC integrates the uncertainty bound information in the control derivation. Controller performances are evaluated based on indices introduced inSection 4.5. Problem is solved using CPLEX 12.2 [50] on a 2.67 GHz machine with 4 GB RAM. Here are the discussions of the results:

4.6.1. Computational aspects

Exploiting the TLDS structure results in the same control law that was obtained from the LTS structure. However, matrix $\mathbf{M}$ of *LTS* has l . m . $r((N(N-1))/2)$ variables (quadratic in$N$) while matrix $\mathbf{M}$ of TLDS has l . m . $r(2N-3)$ variables (linear in N), and hence exhibits shorter computation time. On average, the simulation time for TLDS is 30% less than the LTS structure, as shown in Table 1.

Table 1. Comparison of LTS and TLDS uncertainty feedback parameterizations results for the case of $\delta = 60\%$.

Parametrization	Number of feedback decision variables	Average simulation time for N=24, in (s)	I_e (kWh)	I_d (°Ch)
LTS	$i.m.r\left(\frac{N(N+1)}{2}\right)$	200	16,467	0
TLDS	$3i.m.r(N-1)$	138	16,467	0

4.6.2. Comfort

It is observed from Fig. 13, that the RMPC is the only controller that is able to keep the temperature within the allowed comfort zone, at all times during this test simulation, meaning maintaining minimum level of discomfort $(I_d \leq I_d^*)$, while RBC still does a very good job and MPC fails to do so, resulting to $I_d > I_d^*$ for all $\delta \geq 40$ %. Fig. 14 depicts how*discomfort index Id*, varies with additive model uncertainty δ for MPC, and RMPC. Note that different data points for one δ value refers to simulations with different random sequences. The reason for such a wide variation of the simulation results, specially for large values of δ stems from the fact that depending on the value of the random variable at any time, the resulting disturbance vector can either lead to temperature rise or fall with respect to the nominal disturbance value. It is shown that RMPC manages to keep the perfect comfort level ($Id = 0$), for additive model uncertainty up to $\delta = 75\%$, while the MPC maintains the perfect comfort level for uncertainty bounds up to $\delta = 20\%$. The discomfort index for MPC goes as high as 4.61 °Ch while the value for RMPC reaches 1.2 °Ch in the worst case in the simulations corresponding to $\delta = 100\%$. Since RBC is not a model-based control technique, its performance does not depend on values of δ, hence the straight horizontal line in Fig. 14 ($Id = 0.25$°Ch).

4.6.3. Energy consumption

Fig. 15 depicts the variations of *energy index Ie*, versus the uncertainty bound on the unmodeled dynamics. It is clear that the energy index for RMPC increases dramatically with δ, while the energy index for MPC only changes slightly. However, this comes with the drawback of increased discomfort index for MPC. Fig. 15 also shows energy consumption of RBC ($Ie = 1.43 \times 10^4$ kWh). MPC for all values of δ leads to a lower amount of energy consumption than RBC, but RMPC leads to more energy consumption than RBC soon after $\delta = 35\%$.

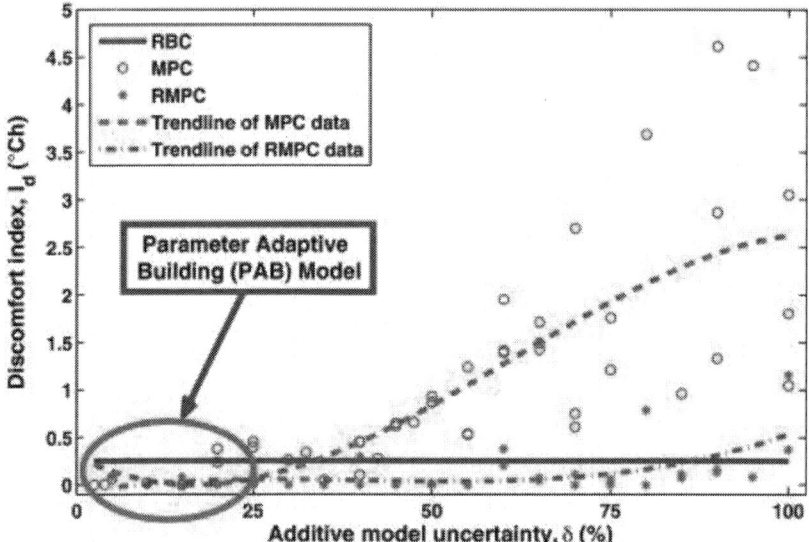

Figure 14. Discomfort index Id versus additive model uncertainty (δ). We generate a uniform random sequence based on the disturbance prediction error value δ. The generated random sequence is used in the simulations for making this graph. Trendlines in this figure are calculated based on least square estimation.

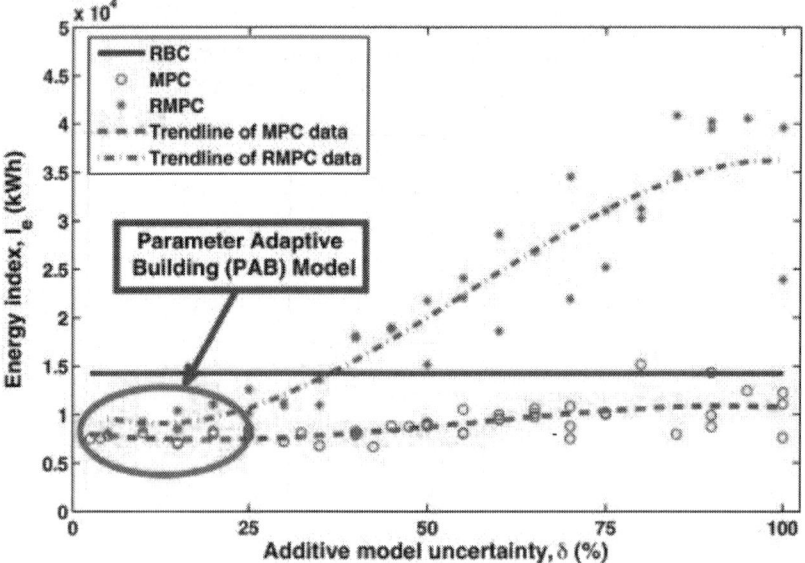

Figure 15. Energy index Ie versus additive model uncertainty (δ). The data points for this graph were generated using a similar technique as in Fig. 14. Trendlines in this figure are calculated based on least square estimation.

4.6.4. Comfort-energy trade-off

An important point to notice from Fig. 15 is how much more energy needs to be supplied to the HVAC system to maintain the comfort level in the presence of imperfect and faulty unmodeled dynamics predictions. Consider the case where $\delta = 75\%$. MPC will lead to a discomfort index of 1.7 °Ch on average, while the RMPC is able to maintain the temperature below a discomfort index of 0.016 °Ch on average. However this level of comfort provided by the RMPC comes at a cost of energy consumption of 3 times more than that of the MPC case. Note that due to the trade-off between comfort and energy consumption, the choice of which controller to use is on the building HVAC operator, and depends on various factors such as criticality of meeting the temperature constraints for the considered thermal zone in the building, and availability and price of energy at that time of the day/year.

As observed from Fig. 14 and Fig. 15 the behavior of controllers vary considerably as the model uncertainty increases. For instance, the energy required to keep the same level of comfort for RMPC in the case of $\delta = 75\%$ is almost 3 times the energy required to provide the same level of comfort when $\delta = 25\%$. Fig. 14 and Fig. 15 show the importance of a good model like PAB in minimizing the energy consumption of building HVAC systems for a desirable comfort level using model-based control techniques by accurately capturing the dynamics of the system.

4.6.5. MPC and RMPC versus RBC

Fig. 16 demonstrates savings of MPC and RMPC versus RBC. As shown, the maximum theoretical energy saving of MPC compared to RBC is 36%, and that of RMPC is 30% for the building studied. These values decrease as model uncertainty increases. Energy saving of MPC versus RBC stays positive even for large values of model uncertainty, while energy saving of RMPC versus RBC is positive only for model uncertainty values up to about 34%, and is negative for larger model uncertainties (i.e. RMPC consumes more energy than RBC).

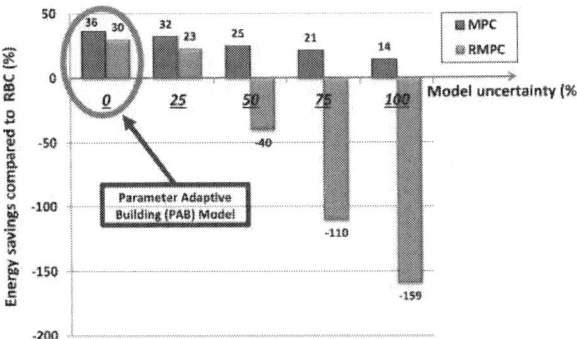

Figure 16. Energy saving of MPC and RMPC compared to RBC as a function of model uncertainty. The blue ellipse shows operating area of the PAB model which keeps the model uncertainty very small. (For interpretation of the references to color in this figure legend, the reader is referred to the web version of the article.)

The results of an extensive study in [51] show that MPC HVAC control can potentially provide 16–41% building energy saving compared to rule-based controllers, which is in agreement with our findings. The saving depends on various factors including climate zone, insulation level, and construction type. Stochastic MPC was shown in [51] to be superior to the rule-based control given the uncertainties in occupancy and weather forecast. Our findings also show that the robust MPC outperforms the rule-based control in terms of energy consumption and user comfort. Although these two MPC techniques (robust and stochastic MPC) both address model uncertainty, they are formulated differently and hence can lead to different performance results. Given the accuracy of the PAB for removing model uncertainty, designing MPC scheme based on PAB is a promising solution for building control problem.

For simulation evaluation of energy consumption and provided comfort level, we have compared the overall performance of the three controllers using *IOP*. The results, as shown in Fig. 17, suggest that for model uncertainties less than 30% MPC is the best controller type. For model uncertainties above 30%, RMPC and RBC are close in performance while for δ between 30% and 67% RMPC is the best, and for model uncertainties larger than 67%, RBC leads to better overall performance than model-based control techniques. This information can be of utility for choosing a controller type for building HVAC system. As described in the paper, proper choice of building HVAC control would depend on the accuracy of the given building model. Range of uncertainties for a given building model can be obtained by taking the difference of the temperature predictions from the building model and temperature measurements from a building. The statistics of such uncertainty can be found once such data is available. The mean and variance of the uncertainty from the statistical analysis can be used to select the best controller type.

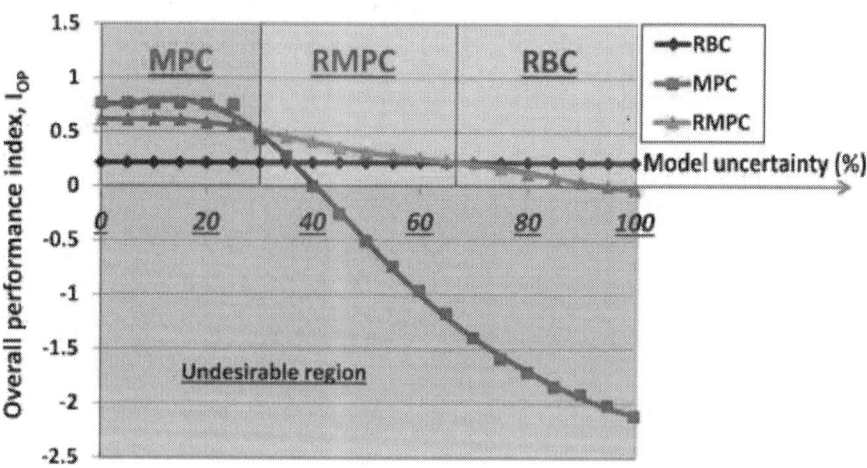

Figure 17. Overall performance index for RBC, MPC and RMPC as a function of model uncertainty. The red zone demonstrates the region which MPC outperforms RMPC and RBC as it yields a higher *IOP*. The green zone

represents the region that *IOP* of RMPC is higher than that of MPC and RBC. RBC dominates in terms of *IOP* in the blue zone. In the gray zone the resulting discomfort index is not acceptable. (For interpretation of the references to color in this figure legend, the reader is referred to the web version of the article.)

5. SUMMARY AND CONCLUSION

Model uncertainty is an unavoidable challenge for modeling and model-based control of a building HVAC system. In this paper, we characterized the impact of model uncertainty on MPC controllers and presented two approaches to minimize model uncertainty for building controls. First, we presented a new modeling framework for simultaneous state estimation and parameter identification of building predictive models. This resulted in a parameter-adaptive building (PAB) model which captures system dynamics through an *online* estimation of time-varying parameters of a building model. The PAB model aims at reducing model uncertainty and can be used for both modeling and control. Second, we presented an MPC framework that is robust against additive uncertainty. The new framework is a closed-loop Robust Model Predictive Control (RMPC) utilizing uncertainty knowledge to enhance the nominal MPC. The RMPC is capable of maintaining the temperature within the comfort zone for model uncertainties up to 75%. The specific findings are listed below:

1. We constructed a nonlinear state space model by augmenting the parameters of the system into the state vector. We exploited the similarities in the physical properties such as wall materials and thicknesses in the building under study, and reduced the number of independent parameters in the building model. A similar approach is expected to apply to other building modeling practices.
2. We presented a PAB modeling framework that uses an unscented Kalman filter (UKF) to simultaneously estimate all the states of the dynamic model and continuously tune the parameters of the building model. The PAB was validated with the experimental HVAC data collected from a building test bed. Successful application of UKF in this work for simultaneous state and parameter estimation of a building model is promising for other building control applications which deal with model uncertainty.
3. We proposed a new uncertainty feedback parameterization of the control input, TLDS, for the closed loop RMPC which results in the same energy and discomfort indices as the previous parameterization, LTS, with a lower number of decision variables, linear in time horizon N, as opposed to quadratic, for the LTS. The new TLDS parameterization results in an average simulation time of 30% less than LTS.
4. Closed loop RMPC outperforms nominal MPC considering the provided level of comfort. However, higher comfort comes at the cost of dramatically higher energy consumption for RMPC. For uncertainty range of 30% to 67%, RMPC leads to better overall performance compared to MPC and

RBC, while it fails to provide a better energy-comfort trade-off if model uncertainty is less than 30% or more than 67%.

5. We proposed a new performance index (*IOP*) to assess buildings' energy consumption and comfort level simultaneously. The *IOP* index is used for evaluating different building controllers. *IOP* index can be used to generate a guideline for choosing appropriate controller type for buildings. This can be helpful for building control community for deciding on a proper controller type based on how accurate an available building model is for model-based controller design.

6. We found that the best choice for controller type changes from MPC to RMPC, and then finally to RBC as the model uncertainty increases. A typical RBC controller outperforms model-based controllers (MPC and RMPC), if building model uncertainty is above 67%.

ACKNOWLEDGEMENTS

Special thanks to Mr. Gregory Kaurala (MTU Energy Management Assistant) for his help in collecting the experimental data, and his technical review. Mehdi Maasoumy is funded by the Republic of Singapore's National Research Foundation through a grant to the Berkeley Education Alliance for Research in Singapore (BEARS) for the Singapore-Berkeley Building Efficiency and Sustainability in the Tropics (SinBerBEST) Program. BEARS has been established by the University of California, Berkeley as a center for intellectual excellence in research and education in Singapore. Alberto Sangiovanni Vincentelli is supported in part by the TerraSwarm Research Center, one of six centers administered by the STARnet phase of the Focus Center Research Program (FCRP), a Semiconductor Research Corporation program sponsored by MARCO and DARPA.

REFERENCES

1. DOE Building Energy Data Book, 2013http://buildingsdatabook.eren.doe.gov/
2. ASHRAE Standard 90.1, American Society of Heating, Refrigerating, and AirConditioning Engineers, 1989.
3. M. Maasoumy, A. Sangiovanni-Vincentelli, Total and peak energy consumption minimization of building HVAC systems using model predictive control, in: IEEE, Design and Test of Computers, 2012.
4. F. Oldewurtel, C.N. Jones, M. Morari, A tractable approximation of chance constrained stochastic MPC based on affine disturbance feedback, in: 47th IEEE Conference on Decision and Control (CDC), 2008, pp. 4731–4736.
5. Y. Ma, F. Borrelli, B. Hencey, A.W. Packard, S. Bortoff, Model predictive control of thermal energy storage in building cooling systems, in: Joint 48th

IEEE Conference on Decision and Control and 28th Chinese Control Conference, Shanghai, PR China, December, 2009.

6. Y. Ma, F. Borrelli, B. Hencey, B. Coffey, S. Bengea, P. Haves, Model predictive control for the operation of building cooling systems, IEEE Transactions on Control Systems Technology 20 (3) (2012) 796–803.

7. F. Oldewurtel, A. Parisio, C.N. Jones, M. Morari, D. Gyalistras, M. Gwerder, V. Stauch, B. Lehmann, K. Wirth, Energy efficient building climate control using stochastic model predictive control and weather predictions, in:IEEE American Control Conference (ACC), 2010, pp. 5100–5105.

8. J. Prí vara, S. Cigler, Z. Vána, ˘ F. Oldewurtel, C. Sagerschnig, E. Zá˘ ceková, ˘ Building modeling as a crucial part for building predictive control, Energy and Buildings 56 (2013) 8–22.

9. M. Maasoumy Haghighi, A. Sangiovanni-Vincentelli, Modeling and optimal control algorithm design for HVAC systems in energy efficient buildings (master thesis), EECS Department, University of California, Berkeley, 2011 http://www.eecs.berkeley. edu/Pubs/TechRpts/2011/ EECS-2011-12.html

10. X. Xu, S. Wang, G. Huang, Robust MPC for temperature control of airconditioning systems concerning on constraints and multiple uncertainties, Building Services Engineering Research and Technology 31 (1) (2010) 39–55.

11. J. Sirok ˘ y, ´ F. Oldewurtel, J. Cigler, S. Prí vara, Experimental analysis of model predictive control for an energy efficient building heating system, Applied Energy 88 (9) (2011) 3079–3087.

12. J. Ma, S.J. Qin, L. Bo, T. Salsbury, Economic model predictive control for building energy systems, in: IEEE PES, Innovative Smart Grid Technologies (ISGT), Detroit, MI, 2011.

13. P.D. Morosan, R. Bourdais, D. Dumur, J. Buisson, Building temperature regulation using a distributed model predictive control, Energy and Buildings 42 (9) (2010) 1445–1452.

14. D. Bertsimas, M. Sim, Tractable approximations to robust conic optimization problems, Mathematical Programming 107 (1/2) (2006) 5–36.

15. Y. Ma, G. Anderson, F. Borrelli, A distributed predictive control approach to building temperature regulation, in: American Control Conference (ACC), San Francisco, CA, 2011.

16. M. Maasoumy, A. Sangiovanni-Vincentelli, Optimal control of HVAC systems in the presence of imperfect predictions, in: ASME, Dynamic System Control Conference (DSCC), Fort Lauderdale, FL, 2012.

17. J. Cigler, S. Prí vara, Z. Vána, ˇ E. Záˇ ceková, ˇ L. Ferkl, Optimization of predicted mean vote index within model predictive control framework: computationally tractable solution, Energy and Buildings 52 (2012) 39–49.

18. S. Cheng, Y. Chen, C.H.K. Chan, T. Lee, H.L. Chan, J. Qin, Q. Zhou, A. Cheung, K. Yu, A robust control strategy for VAV AHU systems and its application, Frontiers in Computer Education 63 (2012) 5–642.

19. G. Huang, F. Jordán, Model-based robust temperature control for VAV airconditioning system, HVAC&R Research 18 (3) (2012) 432–445.

20. H. Moradi, F. Bakhtiari-Nejad, M. Saffar-Avval, Multivariable robust control of an air-handling unit: a comparison between pole-placement and H∞ controllers, Energy Conversion and Management 55 (2012) 136–148.

21. H. Moradi, M. Saffar-Avval, A. Alasty, Nonlinear dynamics, bifurcation and performance analysis of an air-handling unit: disturbance rejection via feedback linearization, Energy and Buildings, Elsevier, 2012.

22. M. Maasoumy, A. Pinto, A. Sangiovanni-Vincenteli, Model-based hierarchical optimal control design for HVAC systems, in: ASME, Dynamic System Control Conference (DSCC), Arlington, VA, 2011.

23. C.Agbi, Z. Song,B.Krogh, Parameter identifiability formulti-zone buildingmodels, in: 51st Annual Conference on Decision and Control (CDC), IEEE, IEEE, 2012, pp. 6951–6956.

24. K. Deng, P.Barooah, P.G.Mehta, S.P.Meyn,Building thermalmodel reductionvia aggregation of states, in: IEEE, American Control Conference (ACC), Baltimore, MD, 2010, pp. 5118–5123.

25. Gregory L. Plett, Sigma-point Kalman filtering for battery management systems of LiPB-based HEV battery packs: part 2: simultaneous state and parameter estimation, Journal of Power Sources 161 (2) (2006) 1369–1384.

26. H. Moradkhani, S. Sorooshian, H.V. Gupta, P.R. Houser, Dual state-parameter estimation of hydrological models using ensemble Kalman filter, Advances in Water Resources 28 (2) (2005) 135–147.

27. M. Rafiee, A. Tinka, J. Thai, A.M. Bayen, Combined state-parameter estimation for shallow water equations., in: IEEE, American Control Conference (ACC), San Francisco, CA, 2011, pp. 1333–1339.

28. P. May-Ostendorp, G.P. Henze, C.D. Corbin, B. Rajagopalan, C. Felsmann, Modelpredictive control of mixed-mode buildings with rule extraction, Building and Environment 46 (2) (2011) 428–437.

29. D.B. Crawley, J.W. Hand, M. Kummert, B.T. Griffith, Contrasting the capabilities of building energy performance simulation programs, Building and Environment 3 (4) (2008) 661–673.

30. F. Oldewurtel, A. Parisio, C.N. Jones, D. Gyalistras, M. Gwerder, V. Stauch, B. Lehmann, M. Morari, Use of model predictive control and

weather forecasts for energy efficient building climate control, Energy and Buildings 45 (2012) 15–27.

31. Y. Agarwal, B. Balaji, R. Gupta, J. Lyles, M. Wei, T. Weng, Occupancy-driven energy management for smart building automation, in: Proceedings of the 2nd ACM Workshop on Embedded Sensing Systems for Energy-efficiency in Building, Zurich, Switzerland, 2010, pp. 1–6.

32. B. Dong, B. Andrews, Sensor-based occupancy behavioral pattern recognition for energy and comfort management in intelligent buildings, in: Proc. of International Building Performance Simulation Association (IBPSA) Conference, Glasgow, Scotland, 2009.

33. M. Maasoumy, Q. Zhu, C. Li, F. Meggers, A. Sangiovanni-Vincentelli, Co-design of control algorithm and embedded platform for HVAC systems, in: The 4th ACM/IEEE International Conference on Cyber-Physical Systems (ICCPS), Philadelphia, PA, 2013.

34. F.P. Incropera, A.S. Lavine, D.P. DeWitt, Fundamentals of Heat and Mass Transfer, John Wiley & Sons Inc., Hoboken, NJ, 2011.

35. Mehdi Maasoumy Haghighi. Controlling energy-efficient buildings in the context of smart grid: a cyber physical system approach. Number UCB/EECS- 2013-244, Dec 2013.

36. C.C. Federspiel, Estimating the inputs of gas transport processes in buildings, IEEE Transactions on Control Systems Technology 5 (5) (1997) 480–489.

37. M.A. Goforth, G.W. Gilcrest, J.D. Sirianni, Cloud effect on thermal downwelling sky radiance,in: Proc. SPIE, Society of Photo-OpticalInstrumentationEngineers, vol. 4710, no. 1, Orlando, FL, 2002, pp. 203–213.

38. National Renewable Energy Laboratory (NREL), 2013 http://www.nrel.gov/

39. S.J. Julier, J.K. Uhlmann, New extension of the Kalman filter to nonlinear systems, in: International Society for Optics and Photonics, AeroSense'97, Seattle, WA, 1997, pp. 182–193.

40. S.J. Julier, J.K. Uhlmann, H.F. Durrant-Whyte, A new approach for filtering nonlinear systems, in: IEEE, Proceedings of the American Control Conference, vol. 3, 1995, pp. 1628–1632.

41. M. Maasoumy, B. Moridian, M. Razmara, M. Shahbakhti, A. SangiovanniVincentelli, Online simultaneous state estimationand parameter adaptationfor building predictive control, in: ASME, Dynamic System and Control Conference (DSCC), Stanford, CA, 2013.

42. ASHRAE Standard. Standard 55-2004, Thermal environmental conditions for human occupancy, 2004.

43. R. de Dear, G.S. Brager, Thermal comfort in naturally ventilated buildings: revisions to ASHRAE Standard 55, Energy and Buildings 34 (6) (2002) 549–561.
44. R. de Dear, G.S. Brager, The adaptive model of thermal comfort and energy conservation in the built environment, International Journal of Biometeorology 45 (2) (2001) 100–108.
45. V. Bradshaw, The building environment: active and passive control systems, 2010.
46. ASHRAE. "Standard 62.1-2004", Ventilation for acceptable indoor air quality, 2004.
47. J. Löfberg, YALMIP: a toolbox for modeling and optimization in MATLAB, in: Proceedings of the CACSD Conference, Taipei, Taiwan, 2004 http://users.isy. liu.se/johanl/yalmip
48. J. Löfberg, Minimax Approaches to Robust Model Predictive Control, Department of Electrical Engineering Linkoping University, 2003.
49. M. Maasoumy, Model predictive control approach to online computation of demand-side flexibility of commercial buildings hvac systems for supply following, IEEE American Control Conference (ACC 2014) (2014).
50. "IBMILOG CPLEX Optimizer",2012http://www.ibm. com/software/ integration/ optimi-zation/cplex-optimizer/
51. D. Gyalistras, M. Gwerder, Use of weather and occupancy forecasts for optimal building climate control (OptiControl): two years progress report, in: Terrestrial Systems Ecology ETH Zurich, Switzerland and Building Technologies Division, Siemens Switzerland Ltd., Zug, Switzerland, 2010.
52. E.A. Wan, R. Van Der Merwe, The unscented Kalman filter, in: Kalman Filtering and Neural Networks, 2001, pp. 221–280.

CHAPTER 8

A Comparison of Energy Consumption Prediction Models Based on Neural Networks of a Bioclimatic Building

Hamid R. Khosravani [1,2], María Del Mar Castilla[3],*, Manuel Berenguel [3], Antonio E. Ruano [1,2] and Pedro M. Ferreira [4]

[1] Faculty of Science and Technology, University of Algarve, Campus Gambelas, Faro, Portugal

[2] Institute of Mechanical Engineering (IDMEC), Instituto Superior Técnico, Universidade de Lisboa, Lisboa, Portugal

[3] Department of Computer Science, Automatic Control, Robotics and Mechatronics Research Group, University of Almería, Agrifood Campus of International Excellence (ceiA3), CIESOL, Joint Center University of Almería-CIEMAT, Almería, Spain

[4] LaSIGE, Faculdade de Ciências, Universidade de Lisboa, Portugal;

ABSTRACT

Energy consumption has been increasing steadily due to globalization and industrialization. Studies have shown that buildings are responsible for the biggest proportion of energy consumption; for example in European Union countries, energy consumption in buildings represents around 40% of the total energy consumption. In order to control energy consumption in buildings, different policies have been proposed, from utilizing bioclimatic architectures to the use of predictive models within control approaches. There are mainly three groups of predictive models including engineering, statistical and artificial intelligence models. Nowadays, artificial intelligence models such as neural networks and support vector machines have also been proposed because of their high potential capabilities of performing accurate nonlinear mappings between inputs and outputs in real environments which are not free of noise. The main objective of this paper is to compare a neural network model which was designed utilizing statistical and analytical methods, with a group of neural network models designed benefiting from a multi objective genetic algorithm. Moreover, the neural network models were compared to a naïve autoregressive baseline model. The models are intended to predict electric power demand at the Solar Energy Research Center (Centro de Investigación en Energía SOLar or CIESOL in Spanish) bioclimatic building located at the University of Almeria, Spain. Experimental results show that the models obtained from the multi objective genetic algorithm (MOGA) perform comparably to the model obtained through a statistical and analytical approach, but they use only 0.8% of data samples and have lower model complexity.

Keywords: predictive model; electric power demand; neural networks; multi objective genetic algorithm (MOGA); data selection

1. INTRODUCTION

Due to fast economic development affected by industrialization and globalization, energy consumption has been steadily increasing over the last years [1,2]. Industry, transportation and buildings are the three main economic sectors which consume a significant amount of energy, with buildings accounting for the biggest proportion. For example in European Union countries, energy consumption in buildings represents about 40% of the total energy consumption [3]. In the USA, more than 44% of domestic energy consumption corresponds to heating, ventilating and air conditioning (HVAC) systems in buildings [4]. Studies have shown that by following the current energy consumption pattern, the world energy consumption may increase more than 50% before 2030 [5], while most of the energy resources are not renewable in nature. Moreover, the usage of energy causes environmental degradation [2]. Therefore, energy consumption management is a very significant problem not only to tackle the losses resulting from increasing consumption patterns but also to improve the performance of building energy systems. With respect to energy management, a variety of policies have been considered. In recent years, there has been a focus on bioclimatic architectures for buildings to reduce the indoor consumption of energy. In this kind of architecture, buildings are designed based on the local climate conditions. These include wind speed and direction, daily exterior temperature and relative humidity, as well as diverse passive solar technologies where heating and cooling techniques passively absorb solar radiation or protect from it without containing mobile elements [6,7,8]. Besides environmental variables, physical properties of buildings are considered in bioclimatic architectures, such as shape, buildings' orientation related to the sun and wind, wall thickness and roof construction [6,9].

Utilizing renewable energy sources such as biomass, hydropower, geothermal, solar, wind and marine energies have been considered as alternatives for conventional energy resources in most developed and developing countries [10,11]. In the European Union, the renewable energies use share is 20% of the total energy consumption and 10% of renewable energies will be used in transportation by 2020 [12]. Using renewable energies not only helps ensure the security of non-renewable energy supply in future, but also minimizes environmental degradation [11].

Prediction of energy use in buildings has received a remarkable amount of attention from researchers [1,3,13,14], as an approach to reduce energy consumption, which is intended to conserve energy and reduce environmental impacts [3]. The prediction of energy usage in buildings and modelling the behaviour of the corresponding energy system, are complicated tasks due to influential factors such as weather variables, building construction, thermal properties of the physical materials and occupants' activities [3]. Furthermore, there are several nonlinear inter-relationships among the involved variables, often in a noisy environment, which amplify the difficulty in identifying the precise interaction among them [15].

The methods aiming to predict building energy consumption can be categorized mainly into statistical, engineering and artificial intelligence ones. A review on prediction methods can be found in [3,16]. Engineering methods, which are detailed comprehensive methods, use the structural properties of buildings in the form of physical principles and thermal dynamics equations, as well as environmental information such as climate conditions, occupants, their activities and HVAC equipment parameters. On the one hand, these methods need a high level of details about the structural and thermal parameters of buildings that are not always available and, on the other hand, since engineering methods depend on complex physical principles, a high level of expertise is needed to elaborately develop the corresponding models [3,17]. To reduce the complexity of the detailed comprehensive engineering methods, simplified methods have been proposed, which can be seen in [18,19].

Statistical methods use historical data to correlate energy consumption as target with most influential variables as inputs. Hence, the quality and quantity of historical data has a crucial role in developing statistical models [17,20]. Unlike engineering methods, statistical methods provide models with a smaller number of variables and much less physical understanding. Regression models, conditional demand analysis (CDA), auto regressive moving average (ARMA), auto regressive integrated moving average (ARIMA) and Gaussian mixture models (GMM) are some instances of statistical models [20,21,22,23].

In recent years, artificial intelligence methods such as neural networks, support vector machines and fuzzy logic have been widely considered in applications of energy consumption. Like statistical methods, artificial intelligence methods use historical data reflecting the behaviour of the process to be modelled. Neural networks have shown a high capability to capture complex nonlinear relationships between inputs and outputs. Since the energy consumption process has a nonlinear behaviour, neural networks are mostly applied in this domain. In addition, they are quicker and easier to develop than engineering and statistical methods, while being accurate estimators. Some instances of neural network based models may be found in [15,17,24,25,26,27,28].

Recently, support vector machines have received much attention as quick methods to build predictive models in energy consumption applications. They can provide models with a high level of generalization based on a number of data. Applications to the prediction of energy utilization can be viewed, for instance, in [29,30,31].

Besides neural network- and support vector machine-based models, another kinds of models which benefit from fuzzy logic have been considered. Fuzzy logic deals with imprecise reality and handles the concept of truth value ranging between completely true and completely false (1–0) [32]. Some models of this type can be seen in [33,34].

As mentioned earlier, both statistical and artificial intelligence methods need sufficient historical data to provide accurate models. In cases where limited amounts of data are available and the information about the process to be modelled is partially known, grey models are suitable alternatives to the prediction of time series associated with processes [35,36,37].

The objective of this paper is to compare a neural network based model obtained in [17] with the models obtained by a multi objective genetic algorithm (MOGA), to predict the electric power demand of the Solar Energy Research Center (Centro de Investigación en Energía SOLar or CIESOL in Spanish) building located at University of Almeria, Spain. The authors in [17] determined the structure and the order of the model by statistical and analytical methods while in this article a non-dominated set of models is generated by a MOGA considering a set of objectives to be optimized. For the sake of completion, the performance of MOGA models is also compared with the results obtained by a naive autoregressive baseline (NAB) approach, introduced in [38].

This paper is organized as follows: in Section 2, the structural properties and power demand profile of the CIESOL building are briefly described. The model proposed in [17] and the models generated by MOGA are widely described in Section 3 and Section 4, respectively. Experimental results are shown in Section 5. Finally, some conclusions are drawn in Section 6.

2. Experimental Setup: The Solar Energy Research Center Building

The CIESOL building (Figure 1a), is a mixed solar energy research centre operated between the Centre for Energy, Environment and Technology (in Spanish the Centro de Investigaciones Energéticas, MedioAmbientales y Tecnológicas—CIEMAT) and the University of Almería, situated in the south-east of Spain. This geographical location is characterized by having a typical semi-desertic Mediterranean climate [39]. This building is divided into two floors with a total surface of approximately 1100 m². More specifically, the upper floor is composed by four laboratories, the director's office and a meeting-room. On the lower floor, five offices, four laboratories, two bathrooms and a kitchen are located. Besides these, the machinery of the solar cooling installation is placed into an environment which occupies two floors.

This building has been designed and built within a research project named PSE-ARFRISOL [40], following bioclimatic architecture criteria. Therefore, it makes a beneficial use of natural ventilation and solar energy in order to reduce energy consumption and CO_2 emissions. To do that, it employs a HVAC system based on solar cooling, which can be observed in Figure 1b, composed by a solar collector field, a hot water storage system, a boiler and an absorption machine with its refrigeration tower [40], and a photovoltaic (PV) power plant with a peak power of 9 kW which provides electricity to the building (Figure 1c,d). Furthermore, a wide network of sensors has been installed in order to monitor the most representative enclosures of the building. Concretely, this network of sensors includes, among others, air temperature, relative humidity, CO_2 concentration, solar radiation, wind velocity and power consumption sensors. Moreover, these sensors are connected to different Compact FieldPoint modules from National Instruments (Madrid, Spain) that are distributed by means of an Industrial Ethernet network all around the building [40]. Data provided by the network of sensors are being stored through a supervisory control and data acquisition (SCADA) system developed with LabVIEW® [40]. Finally, it is necessary to take into account that this building is a research centre which includes chemical, environmental analysis, and modelling and control research groups. Hence, the machinery, other electrical devices and experiments performed by these research groups alter the energy use profile of the building in comparison with more common ones, such as residential buildings.

Figure 1. The Solar Energy Research Center (CIESOL or Centro de Investigación en Energía SOLar in Spanish) building: (**a**) exterior of the CIESOL building; (**b**) solar cooling installation; (**c**) photovoltaic (PV) power plant: PV panels; and (**d**) PV power plant: PV inverters.

2.1. Power Demand Profiles of the Solar Energy Research Center Building

From a power demand point of view, the CIESOL building has some special characteristics mainly derived from the research tasks which are being developed inside it. Therefore, it is necessary to perform an exhaustive analysis of the different energy demand profiles which can be found at the CIESOL building. Specifically, a statistical characterisation involving certain parameters like arithmetic mean ($\bar{x}$), standard deviation (σ), and minimum and maximum values of the power demand (minimum and maximum, respectively) under several conditions (different season and types of days), has been performed (Table 1).

Table 1. Statistical analysis of the power demand profiles (in kW).

Condition	$\bar{x}$	σ	Minimum	Maximum
Working day	24.36	6.39	17.39	44.17
Non-working day	19.45	1.83	12.72	23.86
Winter	26.45	4.55	18.93	39.48
Spring	23.91	6.76	12.56	42.79
Autumn	24.23	4.58	15.85	48.14
Summer	28.74	8.67	16.28	63.48

To predict the power demand within a building, it is necessary to consider numerous energy consuming elements, such as illumination, electrical devices, HVAC systems, *etc.* At the CIESOL building, the element which has the greatest energy consumption is the solar cooling installation. Furthermore, to calculate the total energy demand of the CIESOL building it is necessary to consider both the energy supplied by the electricity company and the energy produced by the PV power plant which is directly consumed by the building, that is, at this moment it is not possible to store the energy from the PV power plant.

Firstly, the main differences according to typical power demand profiles between working and non-working days have been studied, as presented in Figure 2. To do that, a typical day for each demand profile, considering working and non-working days, and each season, has been selected as a function of several environmental variables: mean, maximum and minimum temperature, temperature ranges and solar radiation. The methodology consists of selecting the day with the minimum value obtained from the sum of the weighted absolute difference between each parameter (daily) and the mean value of this parameter along the analysed period. A detailed description of the procedure which has been followed can be found in [41]. It can be observed that power demand on a working day begins to increase around 8:00 am and starts to decrease at 5:00 pm, reaching a stationary value around 8:00 pm, whereas, on a non-working one it has a stationary value approximately equal to 20 kW, mainly due to the machinery and experimental tests performed inside this building. From the perspective of the statistical analysis shown in Table 1, it can be inferred that the mean power demand for a working day is equal to 24.36 kW with a standard deviation of 6.39 kW. On the contrary, for a non-working day, a mean power demand of 19.45 kW and a standard deviation equal to 1.83 kW have been obtained. In addition, working days also present a higher peak power demand, in comparison with non-working days.

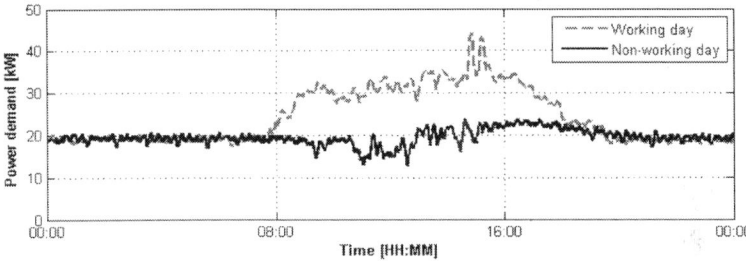

Figure 2. Energy demand profiles for working and non-working days.

Secondly, a detailed examination of the power demand of the CIESOL building through a typical week (from Monday to Sunday), along different environmental conditions has been performed, as shown in Figure 3. The main objectives of this analysis were to determine if there were representative differences among the different seasons of the year and also to identify if there was any characteristic element of the building able to considerably influence its power demand. More specifically, as it can be deduced from Figure 3, the different seasons of the year follow an analogous pattern among working and non-working days. In addition, it can also be inferred that spring and summer seasons present a higher power demand in comparison with winter and autumn. Besides, along the summer season there are several power demand peaks that do not follow any specific pattern associated with the type of day. Therefore, in order to clarify this issue, a detailed analysis of this fact has been performed, and the main conclusions derived from it were that these peaks were associated with the use of a heating pump (for research purposes) and the solar cooling installation. Hence, as the use of both elements is directly associated with the users of the building, it has been decided to take into account the state variables representing these elements within the preliminary list of variables (Table 2). Finally, according to the statistical analysis, it can be concluded that the highest peak power demand and variance is associated with the summer season mainly due to the use the HVAC system for cooling purposes [40].

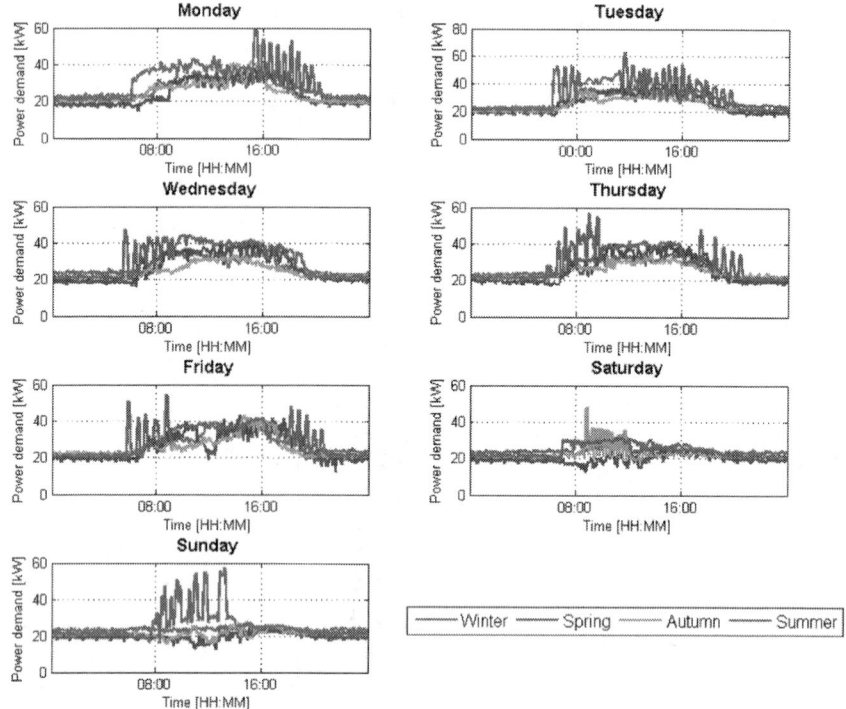

Figure 3. Weekly energy demand profiles for each season.

Table 2. Preliminary list of variables [17].

Variable	Unit	Measurement range
Type of the day (working day/non-working day)	-	(0, 1)
Hour of the day	-	[0,23]
Outdoor temperature	°C	[−5, 50]
Outdoor humidity	%	[0, ..., 100]
Outdoor solar radiation	W/m²	[0,1440]
Outdoor wind speed	m/s	[0,22]
Outdoor wind direction	°	[0,360]
State of the pump B1.1 (off/on)	-	(0, 1)
State of the pump B1.2 (off/on)	-	(0, 1)
State of the pump B2.1 (off/on)	-	(0, 1)
State of the pump B2.2 (off/on)	-	(0, 1)
State of the pump B3.1 (off/on)	-	(0, 1)
State of the pump B3.2 (off/on)	-	(0, 1)
State of the pump B7 (off/on)	-	(0, 1)
State of the boiler (off/on)	-	(0, 1)
State of the absorption machine (off/on)	-	(0, 1)
State of the refrigeration tower (off/on)	-	(0, 1)
State of the heat pump (off/on)	-	(0, 1)
Electric power demand	kW	[0,85]
Electric power injected by the PV plant	kW	[0,9]

Finally, the principal conclusions which have been reached after this precise analysis can be summarized in: (a) there is a clear power demand profile within a week and also, differences between working and non-working day power demand profiles can be undoubtedly established; (b) the power demand for summer is higher mainly due to the typical semi-desertic Mediterranean climate of Almería; and (c) the use of the solar cooling installation has a considerable influence on the final energy consumption.

2.2. Data-Sets Construction

As mentioned previously, in this paper, several energy consumption prediction models based on artificial neural networks (ANN) have been compared. These models have been obtained by means of different methodologies. More specifically, an ANN based prediction model using a MOGA [42,43] has been obtained. Afterwards, this model has been compared with a basic ANN model presented in [17]. To do that, a historic data set acquired at the CIESOL building has been used. Concretely, this data set comprises data from 1 September 2010 to 29 February 2012 with a sample time of 1 min and it includes a preliminary list of variables which can be observed in Table 2. These variables are related with the environmental conditions and the state of the main energy consuming elements of the solar cooling installation.

Subsequently, the selected data set has been split into three different balanced data subsets which have been used to train, test and validate the proposed ANN models. This division has been performed by hand since there were some discontinuities in time series. More information about the methodology followed to obtain these data subsets can be found in [17]. Thereafter, several procedures have been followed in order to obtain different prediction models. A description of these procedures is performed in the following sections.

3. A NON-LINEAR AUTOREGRESSIVE WITH EXOGENOUS INPUTS ARTIFICIAL NEURAL NETWORK MODEL

In [17], a prediction model based on neural networks for the energy consumption of the CIESOL building was proposed. To do that, the neural network Toolbox™ provided by MATLAB® was used. Concretely, the proposed model had a non-linear autoregressive with eXogenous inputs (NARX) architecture, see Equation (1), typified by having a tapped delay line for the input signals set and another one for the output signal, that is, the power demand prediction of the CIESOL building. Moreover, this model has been trained using a gradient-descent based algorithm, more specifically the Levenberg-Marquardt algorithm [44]:

$$y\left[k+1\right] = f\left(u\left[k\right], u\left[k-1\right], \ldots, u\left[k-d_u+1\right]; y\left[k\right], y\left[k-1\right], \ldots, y\left[k-d_y+1\right]\right)$$
$$(1)$$

In the previous equation, u[k] and y[k] represent the input and output signals at time instant k, $d_u \geq 1$, $d_y \geq 1$ (subject to $d_y \geq d_u$) are the memory orders for the input and output tapped delay lines, respectively, and f represents a non-linear mapping function which, in this case, has been approximated by a multilayer perceptron.

Finally, it can be established that the structure of the ANN is completely defined by indicating: (a) the number of hidden layers and the number of neurons in each of them; (b) the number of neurons in the output layer; and (c) the activation function used in each neuron of the hidden and output layers. More specifically, in the model presented in [17], an ANN with only one hidden layer composed by ten neurons with tangent hyperbolic activation functions and one neuron with linear activation function at the output layer has been used, since it is a universal approximator [45].

Afterwards, the selection of input variables from the preliminary variables list, see Table 2, was performed through analytical methods, since they allow to establish the existing linear and non-linear dependencies. Besides, scatter-plots and model tests have been used in order to complete the information provided by analytical methods. A detailed description of these methods can be found in [17]. Therefore, after the application of the methods which have just been mentioned, the preliminary variables list has been reduced to the following ones: type of the day; hour of the day; outdoor temperature and solar radiation; state variables related to the solar cooling installation; and the total power demand of the CIESOL building.

Finally, it is necessary to select the order of the signal inputs, that is, the embedding delay τ and the embedding dimension d [17]. The former has been determined by means of the average mutual information [46], whereas for the latter, optimal values were calculated by the false neighbors method [47]. The list of final input variables and their order can be observed in Table 3.

Table 3. Final list of variables with their order (embedding delay and dimension).

Variable	Unit	Measurement Range	τ	d
Type of the day (working day/non-working day)	-	(0, 1)	1	1
Hour of the day	-	[0,23]	1	1
Outdoor temperature	°C	[−5, 50]	1	4
Outdoor solar radiation	W/m²	[0,1440]	1	4
State of the pump B1.1 (off/on)	-	(0, 1)	1	5
State of the pump B1.2 (off/on)	-	(0, 1)	1	5
State of the pump B2.1 (off/on)	-	(0, 1)	1	5
State of the pump B2.2 (off/on)	-	(0, 1)	1	5
State of the pump B3.1 (off/on)	-	(0, 1)	1	5
State of the pump B3.2 (off/on)	-	(0, 1)	1	5
State of the pump B7 (off/on)	-	(0, 1)	1	5
State of the boiler (off/on)	-	(0, 1)	1	5
State of the absorption machine (off/on)	-	(0, 1)	1	5
State of the refrigeration tower (off/on)	-	(0, 1)	1	5
State of the heat pump (off/on)	-	(0, 1)	1	5
Electric power demand	kW	[0,100]	1	3

4. ARTIFICIAL NEURAL NETWORK BASED MODELS GENERATED BY MULTI OBJECTIVE GENETIC ALGORITHM

MOGA is a design framework implemented in MATLAB®, Python, and C programming languages, which can be applied to determine both the structure and the parameters of ANN based models. In this approach, instead of one model, a non-dominated set of models are generated. From this set, one solution must be selected. In this section the main concepts of MOGA and its application in ANN based models design are addressed. Afterwards, data preparation for MOGA and related experiments are described.

4.1. Multi Objective Genetic Algorithm

In the real world, the optimization of an engineering problem is a complicated task due to the presence of multiple objectives which, most of time, are conflicting with each other, meaning that improving one may deteriorate the other. In this case, there is a Pareto-optimal or non-dominated set in which each solution is not better than the other with respect to the multiple objectives. Figure 4 shows an example of a two objective minimization problem. The whole space of solutions is divided into two groups: the shaded region presents the dominated solutions while the solid curve illustrates the non-dominated set of solutions regarding objectives obj.1 and obj.2. As can be seen in Figure 4, A and B denote two non-dominated solutions.

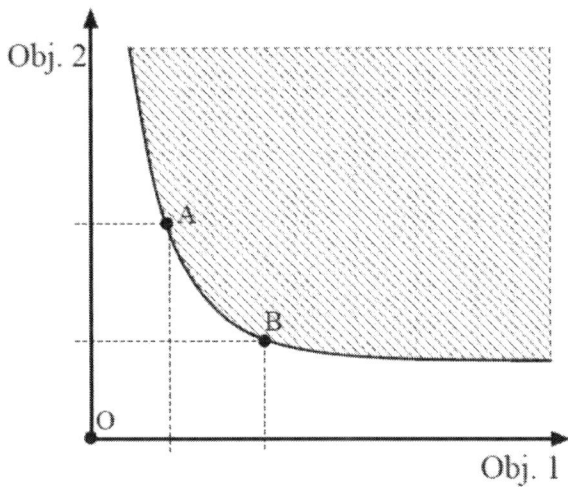

Figure 4. Bi-objective minimization problem. The shaded region presents dominated solutions and solid curve illustrates non-dominated solutions [51].

The goal of a multi-objective optimizer is to improve the surface of non-dominated solutions (*i.e.*, the solid curve) in such a way that the surface approaches the origin (*i.e.*, point 'O' in Figure 4) as much as possible.

Genetic algorithms (GA) are considered promising methods to deal with multi-objective optimization problems [48,49,50]. In MOGA, each individual in the population is evaluated in the space of the multiple objectives rather than in one objective. In addition, at the end of one run of the MOGA, a set of solutions is provided instead of one solution.

Since each individual is evaluated in multi-objective space, the value of objectives should be integrated into a single value in order to assign a fitness to the individual. A simple way is assigning weights to objectives so that each weight reflects the relative importance of its corresponding objective. Afterwards, the summation of the weighted objective's values is considered as a single value to compute and assign a fitness value to the individual. Selecting inappropriate weights leads to wrong searches; additionally, a small variation in weights may result in large changes in objectives.

As a proper alternative, an efficient Pareto-based ranking method has been proposed in [51]. In this way, each individual is ranked based on the number of individuals by which they are dominated. For non-dominated individuals, rank 0 is considered. In most applications, goals and priorities are defined for the objectives so that the Pareto-based ranking method should be modified. For more details about this method, please refer to [51].

4.2. Neural Network Based Model Design by Multi Objective Genetic Algorithm

The problem of designing a neural network based model can be divided into two sub-problems as follows:

Neural network structure: the network inputs and the number of hidden layers/neurons in the network;

Neural network parameters: they depend on the model chosen and are usually determined by a suitable training algorithm.

In this case a radial-basis function (RBF) Neural Network (NN) will be used. The output of a RBF model is given by:

$$o[k] = w_{l+1} + \sum_{j}^{l} w_j e^{-\frac{\| i[k] - \mathbf{C}(j) \|_2^2}{2\sigma_j^2}} \tag{2}$$

In Equation (2), o[k] denotes the output, at instant k, ij[k] is the jth input at that instant, w represents the vector of the linear weights, C(j) represents the vector (extracted from the $\mathbf{C}$ matrix) of the centers associated with the hidden neuron j, σj is its spread, and $\|\|2$ represents the Euclidean distance. The network parameters, which will be denoted as the parameter vector p, are therefore $\mathbf{C}$, σ and w.

According to the above sub-problems, in order to design a neural network based model that satisfies a set of defined goals, it is necessary to define a set of quality measures in the form of objectives for each sub-problem.

Assume that D=(X,y) is a data set composed of N input-output pairs, which is divided into a training set, Dt, a generalization or testing set Dg, and a validation set Dv. Assume also that F is a set of all possible input features (delayed values of the modelled and exogenous variables) and p is the parameter vector. The problem of designing a neural network based model by MOGA can be expressed as follows:

Dataset D, the range d∈[dm, dM] of input features from F and the range n∈[nm, nM] of hidden neurons are given to the MOGA. After executing, the MOGA generates a non-dominated set of RBF models that minimize [μp,μs] where μp and μs denote a set of objectives related to the neural network parameters p and the neural network structure, respectively.

In our work, the corresponding objectives for μp and μs were considered as follows:

$$\mu_p = \left[\varepsilon\left(D^t\right), \; \varepsilon\left(D^g\right), \varepsilon\left(D^s, PH\right) \right] \tag{3}$$

$$\mu_s = \left[O\left(\mu\right) \right] \tag{4}$$

where ε(Dt) and ε(Dg) denote the root-mean-square errors (RMSE) of the training set Dt and the testing set Dg, respectively. Consider a given prediction horizon (*PH*) and a simulation set Ds (with *m* consecutive input-output pairs). Assuming that E(Ds,PH) is an error matrix:

$$\mathbf{E}\left(D^s, PH\right) = \begin{bmatrix} e\left[1,1\right] & e\left[1,2\right] & \cdots & e\left[1, PH\right] \\ e\left[2,1\right] & e\left[2,2\right] & \cdots & e\left[2, PH\right] \\ \vdots & \vdots & \ddots & \vdots \\ e\left[m-PH,1\right] & e\left[m-PH,1\right] & \cdots & e\left[m-PH, PH\right] \end{bmatrix}$$

$$\tag{5}$$

where e[i,j] is the model prediction error taken from instant *i* of Ds at step *j* within the *PH*. By denoting ρ(.,i)as the RMS function operating over the *i*th column of its argument matrix, then ε(Ds,PH) is defined as:

$$\varepsilon\left(D^s, PH\right) = \sum_{i=1}^{PH} \rho\left(\mathbf{E}\left(D^s, PH\right), i\right) \tag{6}$$

O(μ) denotes the model complexity, which is equal to the number of input features + 1, multiplied by the number of hidden neurons, reflecting the RBF input-output topology.

The MOGA searches the space spanned by the number of neurons and the input features, *i.e.*, the model structure. Each individual in the population has a chromosome representation consisting of two components. The first corresponds to the number of hidden neurons The second component is a string of integers, each one representing the index of a particular feature in F. The chromosome representation is shown in Figure 5.

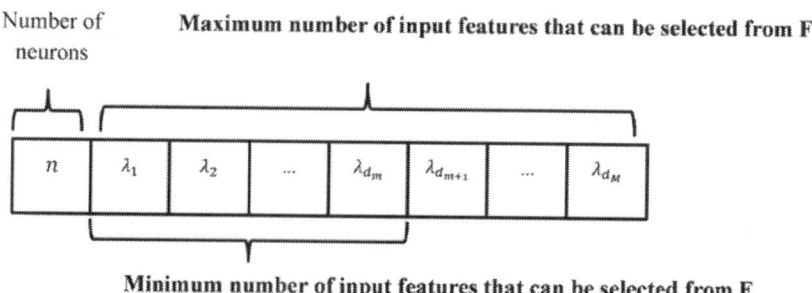

Figure 5. Chromosome representation.

Before being evaluated in the MOGA, each model has its parameters determined by a Levenberg-Marquardt algorithm [37] minimizing an error criterion that exploits the linear-nonlinear relationship of the RBF NN model parameters [52,53,54]. The initial values of the nonlinear parameters (C and σ) are chosen randomly, or with the use of a clustering algorithm, w is determined as a linear least-squares solution, and the procedure is terminated using the early-stopping approach [55] within a maximum number of iterations.

4.3. Model Design Cycle

Briefly, the model design optimization problem is a sequence of actions which are undertaken by the model designer. These actions are repeated until the pre-specified design goals are achieved. There are three main actions in model design cycle: problem definition, solution(s) generation and analysis of results. In the problem definition stage, the datasets, the ranges of features and neurons are defined, as well as the objectives. After this stage, the MOGA does a guided search to obtain models that satisfy the predefined objectives and goals. In the third stage, the set of models obtained that lie in the Pareto front are analyzed. In this set, the performance of the models in the validation set (not involved in the design) is of paramount importance. If good solutions are found, the process stops. Otherwise, based on the results analysis, the search space can be reduced, and/or the objectives and goals can be redefined, therefore restricting the trade-off surface coverage. A more detailed description of the MOGA based ANN design framework can be found, for instance, in [43].

4.4. Data Preparation

After an analysis of the original data, a new code was considered for the feature "day type". The new code refers to "special days". By comparing the

amount of energy consumption for working and non-working days, it has
been revealed that for some days over the years 2010 and 2011, the amount of
energy consumption has an average value between working and non-working
days. By comparing these special days with the Spanish calendar for both
years, it was found that those days occurred in the early days of the year, or
in working days which were located between national/regional holidays and
weekends. Based on that, these special days received the code 0.5. Figure
6 shows the distribution of whole data samples in terms of "day type".

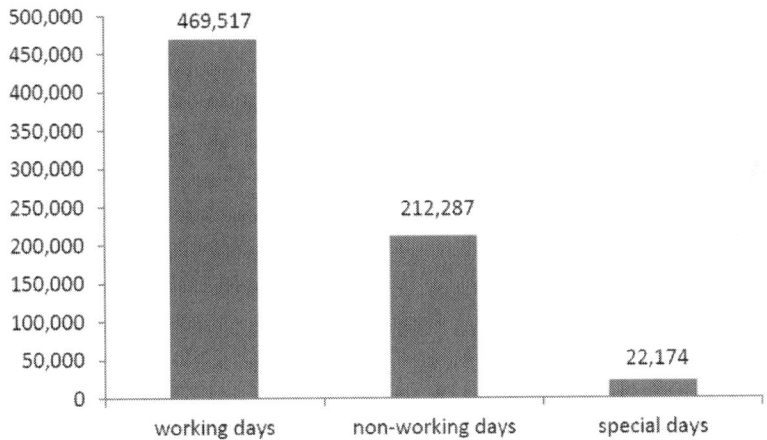

Figure 6. Distribution of original data samples in terms of day type from 1
September 2010 to 29 February 2012.

Since the original data was obtained with a sampling interval of 1 min, its
size was too large (514,762 samples) to be handled by the MOGA
framework, and was reduced in several stages. Due to presence of gaps in the
data, there were 51 consecutive periods over the whole data. In the first stage,
each period was divided into one week length segments. Based on these
divisions, those durations whose length was less than two weeks were
ignored in this work. This stage resulted into 13 periods containing at least
two weeks of data. Table 4 shows the periods selected in the first stage.

In the second stage, the data for all periods was reduced by a factor of 15
by averaging every 15 consecutive samples inside each segment. The
sampling interval was then increased to 15 min.

In the third stage, by starting from the second week within each period,
three random days along with the last seven consecutive days were selected
as lags for each variable. This way, a data set D with 8640 samples was
obtained. Figure 7 shows the distribution of samples of data set D in terms of
"day type".

Table 4. The periods selected in the first stage.

Period Number	Start	End
1	2 September 2010 00:00:00	15 September 2010 23:59:00
2	24 September 2010 00:00:00	14 October 2010 23:59:00
3	9 November 2010 00:00:00	22 November 2010 23:59:00
4	27 December 2010 00:00:00	9 January 2011 23:59:00
5	11 January 2011 00:00:00	31 January 2011 23:59:00
6	9 February 2011 00:00:00	1 March 2011 23:59:00
7	11 March 2011 00:00:00	31 March 2011 23:59:00
8	2 June 2011 00:00:00	22 June 2011 23:59:00
9	8 July 2011 00:00:00	1 September 2011 23:59:00
10	14 October 2011 00:00:00	27 October 2011 23:59:00
11	5 November 2011 00:00:00	23 December 2011 23:59:00
12	29 December 2011 00:00:00	11 January 2012 23:59:00
13	19 January 2012 00:00:00	8 February 2012 23:59:00

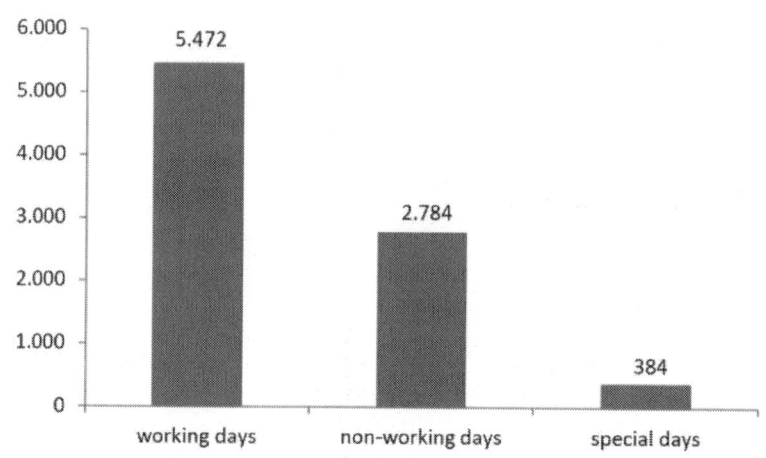

Figure 7. Distribution of samples in data set D in terms of day type.

4.5. Design Experiments

Based on the model design cycle described in Section 4.3, several designs were conducted in such a way that their results led to the definition of a new design, by redefining variables and their corresponding lag terms, as well as imposing restrictions on objectives.

In a first step, we conducted designs with features requiring lag terms spread over at most 7 days. After analyzing and comparing the results with those obtained in [17], the spread of lags was reduced to cover at most 2 days, and finally to cover at most one day. Based on that, four new designs were carried out.

For all designs, data set D, stated in Section 4.4, containing 8640 samples was used. Since a sampling interval of 15 min was used, and the objective was to obtain forecasts of electric power 1 h-ahead, a prediction horizon of four steps was employed. In this work, as in [17], two groups of models were considered. The first group contains simple models where only weather variables are used as exogenous variables. The second group considers complete models involving both weather and solar cooling operation variables. The list of candidate variables used and the range of lags for the design experiments are given in Table 5 and Table 6, respectively.

Table 5. List of variables used.

Variable	Notation	Unit	Range in D
Electric power demand added up with the electric power supplied by the PV plant	$x1$	kW	[11.73, 74.65]
Day type (working day/non-working day/semi-holidays)	$x2$	-	(0, 0.5, 1)
Outdoor temperature	$x3$	°C	[2.73, 43.79]
Outdoor solar radiation	$x4$	W/m^2	[0, 1127.81]
State of pump B1.1 (off/on)	$x5$	-	(0, 1)
State of Pump B1.2 (off/on)	$x6$	-	(0, 1)
State of Pump B2.1 (off/on)	$x7$	-	(0, 1)
State of Pump B2.2 (off/on)	$x8$	-	(0, 1)
State of Pump B7 (off/on)	$x9$	-	(0, 1)
State of the boiler (off/on)	$x10$	-	(0, 1)
State of the absorption machine (off/on)	$x11$	-	(0, 1)
State of the cooling tower (off/on)	$x12$	-	(0, 1)
State of the heat pump (off/on)	$x13$	-	(0, 1)

Table 6. Description of the lags used.

Variable	Experiment I	Experiment II	Experiment III	Experiment IV
$x1$	20 lags over 1 day	20 lags over 1 day	20 lags over 1 day	20 lags over 1 day
$x2$	0 lags	0 lags	0 lags	0 lags
$x3$	20 lags over 1 day	20 lags over 1 day	20 lags over 1 day	20 lags over 1 day
$x4$	20 lags over 1 day	20 lags over 1 day	20 lags over 1 day	20 lags over 1 day
$x5$	-	-	1 lag	1 lag
$x6$	-	-	1 lag	1 lag
$x7$	-	-	1 lag	1 lag
$x8$	-	-	1 lag	1 lag
$x9$	-	-	1 lag	1 lag
$x10$	-	-	1 lag	1 lag
$x11$	-	-	1 lag	1 lag
$x12$	-	-	1 lag	1 lag
$x13$	-	-	1 lag	1 lag

As it can be seen in Table 6, Experiments I and II correspond to simple models in which only weather variables have been used; Experiments III and IV consider complete models. In Table 6, "lag 0" for variable "day type" ($x2$) is translated into the day type of instant k+1 for which the electric power demand is predicted. In fact, weather and electric power demand variables are strongly related to their most recent values and also, to a certain extent, to their values 24 h before. As a result, for $x1$, $x3$ and $x4$ a heuristic, proposed in [43], was used to select 20 lags over one full day, in such a way that more recent values predominate in the set of searchable lags for these variables. Hence, based on this heuristic, the 20 lags used are [1–7, 9, 11, 13, 16, 20, 24, 29, 36, 43, 53, 65, 79, 96]. In this list, and as an example, lags 1 and 2 denote delays of 15 and 30 min, respectively. The objectives and the corresponding goals are given in Table 7.

Table 7. Objectives and their corresponding restriction of experiments.

Objectives	Experiment I	Experiment II	Experiment III	Experiment IV
$\varepsilon\left(D^{t}\right)$	Minimize	<0.059	Minimize	<0.054
$\varepsilon\left(D^{g}\right)$	Minimize	<0.061	Minimize	<0.052
$\varepsilon\left(D^{s},16\right)$	Minimize	Minimize	Minimize	Minimize
$O\left(\mu\right)$	Minimize	<317	Minimize	<444

Regarding MOGA's design framework parameters specification, for experiments I and III, the range [dm, dM], where dm and dM are the minimum and maximum number of features, was set to [1, 30] while for experiments II and IV they were set to [1, 15] and [1, 21], respectively. Similarly, for experiments I and III, the range [nm, nM], where nm and nM are the minimum and maximum number of neurons, was set to [2,30] while for experiments II and IV, these ranges were set to [1, 18] and [1, 21], repectively. For all designs, the population size and the number of generations were set to 100. For each experiment, a proper sub dataset DW was derived from data set D whose features are those columns of D which correspond to the lags defined in the corresponding experiment.

In order to generate training, testing and validation sets for each experiment, firstly the ApproxHull algorithm [56] was applied on corresponding DW to obtain convex points reflecting the whole input range in which the model is supposed to be used. Secondly, 50% of whole samples in DW were used to generate training, testing and validation sets with proportions of 60%, 20% and 20%, respectively. In this step all convex points were incorporated in the training set. Afterwards, the remaining samples were shared randomly into the rest of the training set, and the testing and validation sets. Regarding the simulation dataset Ds, 1344 consecutive samples from 1 October 2010 00:00:00 to 14 October 2010 23:59:00 were considered. In this set, the rows correspond to the variables used, whose samples are in each column while, for the other sets, the number of rows correspond to the patterns, and the number of columns to the features. The size of training, testing and validation datasets as well as the simulation dataset of each experiment is given in Table 8.

After one run of the MOGA for each experiment, the non-dominated and preferred sets of models were generated. In the case that no restriction is considered on objectives, the non-dominated set is the same as preferred set;

otherwise, the preferred set is a subset of the non-dominated set whose solutions satisfy the goals. Please refer to [51] for further information about how the preferred set can be obtained from the non-dominated set by applying the preferably criterion. The number of models in non-dominated and preferred sets for each experiment is given in Table 9.

Table 8. Size of training, testing and validation sets.

Data set	Experiment I	Experiment II	Experiment III	Experiment IV
D^t	2592 × 62	2592 × 62	2592 × 71	2592 × 71
D^g	864 × 62	864 × 62	864 × 71	864 × 71
D^v	864 × 62	864 × 62	864 × 71	864 × 71
D^s	4 × 1344	4 × 1344	13 × 1344	13 × 1344

Table 9. Size of non-dominated and preferred sets.

Experiment	Non-dominated set	Preferred set
Experiment I	346	346
Experiment II	238	88
Experiment III	289	289
Experiment IV	366	182

RESULTS AND DISCUSSION

The models presented in this paper have been tested and compared by means of real data acquired at the CIESOL building. To do that, a battery of tests has been selected according to certain representative characteristics, such as, the type of day (working and non-working days), the season of the year and the quantity of solar radiation (sunny and cloudy days). A complete description of the battery of tests is shown in Table 10. Furthermore, a prediction horizon over 1 h has been set mainly due to the energy price changes and the dynamic behaviour of indoor temperature [17].

Since in MOGA related experiments, the data used a sampling interval of 15 min, each test in Table 10 contains 96 samples. Moreover, the corresponding prediction horizon over 1 h is equal to four steps. For the model proposed in [17], each test includes 1440 samples due to the 1 min sampling rate. Hence, the corresponding prediction horizon over 1 h is equal to 60 steps. For convenience, the complete model proposed in [17] and the models obtained by MOGA will be denoted as PREVIOUS and MOGA models, respectively. In order to compare the MOGA models obtained from each experiment with the PREVIOUS model, one model was selected from the non-dominated/preferred set, with a good compromise between performance and complexity.

Table 10. Battery of tests performed.

Test	Day	Temperature	Radiation	Date
(A)	Working day	Summer	Sunny	29 June 2011
(B)	Non-working day	Summer	Sunny	19 September 2010
(C)	Working day	Winter	Cloudy	15 February2011
(D)	Non-working day	Winter	Sunny	20 February 2011
(E)	Non-working day	Winter	Cloudy	28 February 2011
(F)	Non-working day	Summer	Cloudy	2 July 2011

In our work, models I–IV were the selected MOGA models from experiments I–IV, respectively. Information about the selected MOGA models as well as the PREVIOUS is given in Table 11. Using the notation of Table 6, the formal description of models I–IV is given by Equations (7)–(10), respectively:

$$\hat{y}(k+1) = f_1(x_1(k), \ldots, x_1(k-6), x_1(k-8), x_1(k-11), x_1(k-12), x_1(k-19), x_2(k+1), \\ x_3(k-2), x_3(k-7), x_3(k-10), x_4(k-4), x_4(k-10), x_4(k-17)) \tag{7}$$

$$\hat{y}(k+1) = f_2(x_1(k), \ldots, x_1(k-4), x_1(k-6), x_1(k-9), x_1(k-10), x_1(k-15), \\ x_1(k-18), x_3(k-9), x_4(k), x_4(k-8), x_4(k-18)) \tag{8}$$

$$\hat{y}(k+1) = f_3(x_1(k-1), x_1(k-3), x_1(k-4), x_1(k-5), x_1(k-7), x_1(k-10), x_1(k-11), \\ x_1(k-12), x_1(k-14), x_1(k-15), x_1(k-16), x_2(k+1), x_3(k), x_3(k-2), \\ x_3(k-3), x_3(k-4), x_3(k-8), x_3(k-12), x_3(k-13), x_3(k-15), x_3(k-16), \\ x_4(k-2), x_4(k-3), x_4(k-5), x_4(k-7), x_4(k-12), x_7(k), x_{11}(k), x_{13}(k)) \tag{9}$$

$$\hat{y}(k+1) = f_4(x_1(k), x_1(k-1), x_1(k-2), x_1(k-3), x_1(k-5), x_1(k-17), \\ x_2(k+1), x_3(k-18), x_4(k-3), x_4(k-5), x_4(k-10), x_4(k-14), \\ x_4(k-15), x_4(k-18), x_9(k), x_{10}(k), x_{11}(k), x_{13}(k)) \tag{10}$$

$\hat{y}(k+1)$ in Equations (7)–(10) is the output of the corresponding RBF neural network, representing o[k] in Equation (2). Each function fj, {j=1,2,3,4} has its own set of input terms. These input terms, all together, constitute the input data sample at instant k corresponding to i[k] in Equation (2).

Table 11. Selected multi objective genetic algorithm (MOGA) models and PREVIOUS model. Artificial neural networks: ANN.

Model	Number of features	Number of neurons	Complexity
Model I	18	13	247
Model II	14	18	270
Model III	29	11	330
Model IV	18	20	380
NARX-ANN	67	10	680

To compare MOGA models with the PREVIOUS model over the battery of tests stated in Table 10, five statistical criteria were considered: mean absolute error (*MAE*), mean relative error (*MRE*), Mean absolute percentage error (*MAPE*), maximum absolute error (*MaxAE*) and standard deviation of predicted values (σ). These criteria can be calculated according to Equations (11)–(15):

$$MAE = \frac{1}{N} \sum_{i=1}^{N} |x(i) - \hat{x}(i)| \tag{11}$$

$$MRE = \frac{1}{N} \sum_{i=1}^{N} \frac{|x(i) - \hat{x}(i)|}{x(i)} \tag{12}$$

$$MAPE = \frac{100\%}{N} \sum_{i=1}^{N} \frac{|x(i) - \hat{x}(i)|}{|x(i)|} \tag{13}$$

$$MaxAE = \max(AE(x, \hat{x})) \tag{14}$$

$$\sigma = \sqrt{\frac{1}{N} \sum_{i=1}^{N} (\hat{x}(i) - \bar{\hat{x}}(i))^2} \tag{15}$$

In Equations (11)–(15), N, x and x^ denote the number of samples, measured values and predicted values of the variable, respectively. The evaluations of MOGA and PREVIOUS models over the battery of tests for a prediction horizon of 1 h are given in Table 12, Table 13, Table 14, Table 15, Table 16 and Table 17. The best values for each criterion are identified in bold.

Regarding test A, a working sunny day in summer, Model I, as a simple model, not only has minimum values in terms of *MAE*, *MRE* and *MAPE* among other MOGA models but also has a better performance than PREVIOUS in terms of these criteria. In this test, in overall, simple models I and II have better performance in comparison with complete models III and IV.

Table 12. Results obtained by MOGA and PREVIOUS models over Test A, for a prediction horizon (*PH*) of 1 h. Mean absolute error (*MAE*); mean relative error (*MRE*); mean absolute percentage error (*MAPE*); maximum absolute error (Max*AE*).

Parameter	Model I	Model II	Model III	Model IV	PREVIOUS
MAE (kW)	**1.92**	2.14	2.28	3.55	1.96
MRE (kW)	**0.06**	0.08	0.07	0.12	**0.06**
MAPE (%)	**6.29**	8.11	7.66	12.39	6.38
Max*AE* (kW)	12.36	14.22	**10.21**	13.82	10.99
σ (kW)	8.92	7.86	8.91	**6.99**	7.17

Table 13. Results obtained by MOGA and PREVIOUS models over Test B, for a *PH* of 1 h.

Parameter	Model I	Model II	Model III	Model IV	PREVIOUS
MAE (kW)	0.95	1.22	1.29	0.93	**0.84**
MRE (kW)	0.05	0.07	0.07	**0.05**	0.05
MAPE (%)	5.60	7.21	7.86	5.80	**5.13**
Max*AE* (kW)	3.60	**3.15**	4.83	3.38	3.59
σ (kW)	2.01	2.48	1.78	1.75	**1.52**

Table 14. Results obtained by MOGA and PREVIOUS models over Test C, for a *PH* of 1 h.

Parameter	Model I	Model II	Model III	Model IV	PREVIOUS
MAE (kW)	1.99	3.46	**1.75**	1.95	1.86
MRE (kW)	0.06	0.1	**0.06**	0.06	0.06
MAPE (%)	6.62	10.55	**6.25**	6.40	6.26
Max*AE* (kW)	8.82	16.56	**5.69**	7.04	8.15
σ (kW)	**6.04**	6.94	6.78	7.75	6.70

Table 15. Results obtained by MOGA and PREVIOUS models over Test D, for a *PH* of 1 h.

Parameter	Model I	Model II	Model III	Model IV	PREVIOUS
MAE (kW)	0.94	1.12	**0.82**	0.88	1.08
MRE (kW)	0.04	0.05	**0.03**	0.04	0.05
MAPE (%)	4.21	5.34	**3.81**	4.17	4.86
Max*AE* (kW)	**4.65**	6.35	5.20	5.45	6.28
σ (kW)	1.95	1.64	**1.08**	1.72	1.52

Table 16. Results obtained by MOGA and PREVIOUS models over Test E, for a *PH* of 1 h.

Parameter	Model I	Model II	Model III	Model IV	PREVIOUS
MAE (kW)	1.38	1.45	**1.16**	1.30	1.49
MRE (kW)	0.06	0.06	**0.05**	**0.05**	0.06
MAPE (%)	6.00	6.30	**5.06**	5.77	6.38
Max*AE* (kW)	**4.44**	5.59	4.81	4.49	6.89
σ (kW)	1.80	1.39	**1.28**	1.65	1.43

Table 17. Results obtained by MOGA and PREVIOUS models over Test F, for a *PH* of 1 h.

Parameter	Model I	Model II	Model III	Model IV	PREVIOUS
MAE (kW)	1.02	**0.80**	1.35	0.89	0.95
MRE (kW)	0.04	**0.03**	0.06	0.04	0.04
MAPE (%)	4.87	**3.70**	6.53	4.28	4.31
Max*AE* (kW)	3.43	**2.63**	5.68	3.73	3.75
σ (kW)	1.95	1.99	**1.44**	1.97	1.88

With respect to test B, a non-working sunny day in summer, Model IV, as a complete model, has minimum values of *MAE*, *MRE* and σ in comparison with other MOGA models; with respect to *MaxAE*, it has a compromise performance between Model II and PREVIOUS.

In test C, a working cloudy day in winter, and in test D, a non-working sunny day in winter, the complete model III has minimum values in terms of *MAE*, *MAPE* and Max*AE* among all models. Model I, a simple model, has also a good performance; actually better in 4 criteria than the complete PREVIOUS model, in test D.

In test E, a non-working cloudy day in winter, both simple and complete MOGA models have lower values in terms of *MAE*, *MAPE* and Max*AE* than the PREVIOUS model. Model III has better performance in all criteria.

Regarding test F, a non-working cloudy day in summer, simple model II and complete model IV have better performance in terms of *MAE*, *MAPE* and *MaxAE* than PREVIOUS model. In this comparison, model II has minimum values in all criteria, except σ.

According to Table 12, Table 13, Table 14, Table 15, Table 16 and Table 17, in the group of simple models, model I, in most cases, has better performance than model II. In the group of complete models, model III, in most cases, is better than model IV. Figure 8, Figure 9 and Figure 10 show the comparison between measured and predicted value of electric power demand in CIESOL building, over tests A–F for a prediction horizon of 1 h, for the PREVIOUS model, model I and III, respectively.

Comparing the performance of all MOGA models over the battery of tests, in general complete models III and IV have a better performance in winter than in summer, while simple model I has a compromise performance between summer and winter.

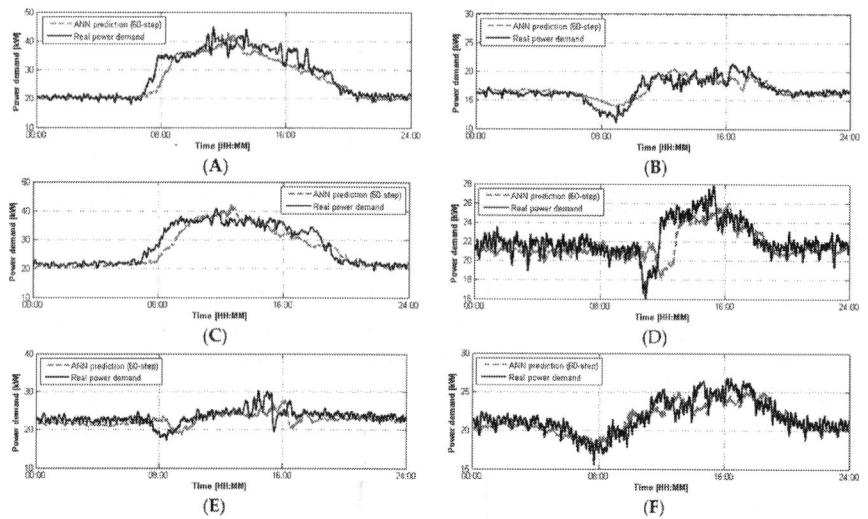

Figure 8. Prediction results for tests (**A–F**) using the PREVIOUS model.

5.1. Comparison of Multi Objective Genetic Algorithm Models with Naive Autoregressive Baseline Approach

The performance of MOGA models was also compared with a NAB model, introduced in [38]. The NAB approach considers, as estimate of the electric power demand at instant k, the measured value of consumption at the corresponding instant of time, in the same day of the previous week. It is therefore a simple model which does not need any computation to predict electric power demand at each time instant k. To apply the NAB approach to tests A–F, consecutive data corresponding to the previous week would be needed. Since there were several gaps in the whole dataset among tests A–F, only for tests D and E, corresponding to special days in winter, consecutive data exist to implement this method. In order to evaluate the NAB model in summer, we considered another special day in summer, corresponding to 6 August 2011, hereinafter called test G. For convenience, the description of the tests D, E and G is given in Table 18.

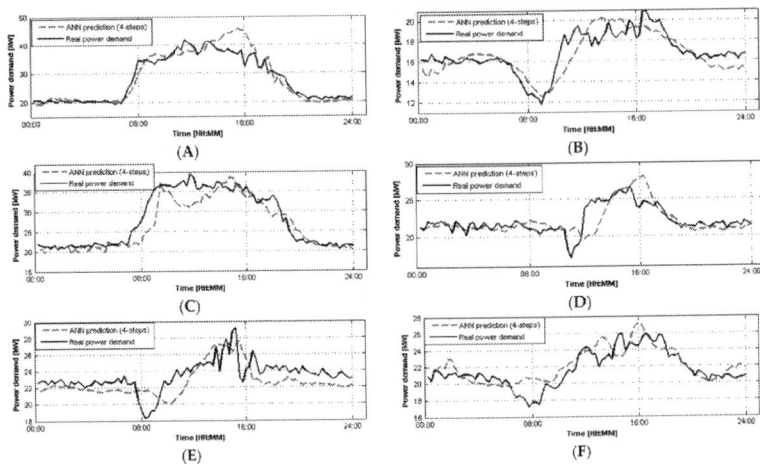

Figure 9. Prediction results for tests (**A–F**) using model I.

In order to compare the performance of NAB model with MOGA models and PREVIOUS model, the three models were evaluated over the battery of tests stated in Table 18. The results obtained over tests D, E and G are given in Table 19, Table 20 and Table 21. Please note that the results of MOGA models and PREVIOUS model, for tests D and E, are obtained from Table 14 and Table 15, respectively, and are reproduced here for easy of comparison with the NAB approach.

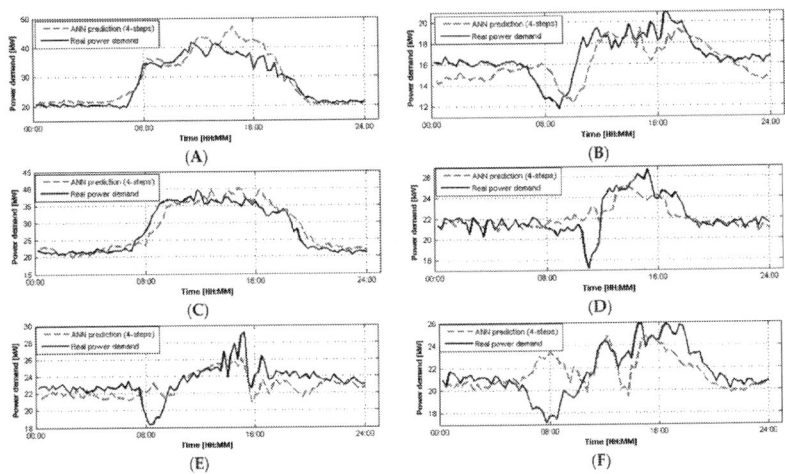

Figure 10. Prediction results for tests (**A–F**) using model III.

Table 18. Battery of tests performed to compare the naive autoregressive baseline (NAB) model with the neural networks models.

Test	Day	Temperature	Radiation	Date
(D)	Non-working day	Winter	Sunny	20 February 2011
(E)	Non-working day	Winter	Cloudy	28 February 2011
(G)	Non-working day	Summer	Sunny	6 August 2011

Table 19. Results obtained by neural network and NAB models over Test D, for a *PH* of 1 h.

Parameters	Model I	Model II	Model III	Model IV	PREVIOUS	NAB
MAE (kW)	0.94	1.12	**0.82**	0.88	1.08	1.9439
MRE (kW)	0.04	0.05	**0.03**	0.04	0.05	0.0856
MAPE (%)	4.21	5.34	**3.81**	4.17	4.86	8.5575
MaxAE (kW)	**4.65**	6.35	5.20	5.45	6.28	6.8341
σ (kW)	1.95	1.64	**1.08**	1.72	1.52	1.8933

Table 20. Results obtained by neural network and NAB models over Test E, for a *PH* of 1 h.

Parameters	Model I	Model II	Model III	Model IV	PREVIOUS	NAB
MAE (kW)	1.38	1.45	**1.16**	1.30	1.49	4.8314
MRE (kW)	0.06	0.06	**0.05**	0.05	0.06	0.2086
MAPE (%)	6.00	6.30	**5.06**	5.77	6.38	20.8610
MaxAE (kW)	**4.44**	5.59	4.81	4.49	6.89	13.0946
σ (kW)	1.80	1.39	**1.28**	1.65	1.43	5.6539

Table 21. Results obtained by neural network and NAB models over Test G, for a *PH* of 1 h.

Parameters	Model I	Model II	Model III	Model IV	PREVIOUS	NAB
MAE (kW)	0.8297	1.2684	0.8089	**0.7598**	0.7787	3.2966
MRE (kW)	0.0472	0.0745	0.0465	0.0434	**0.0432**	0.1909
MAPE (%)	4.7154	7.4521	4.648	4.3363	**4.3154**	19.0867
MaxAE (kW)	3.7347	7.4701	4.7188	3.8647	**2.9473**	13.6549
σ (kW)	2.08	2.216	1.5135	**1.2575**	1.822	3.8805

Regarding these tests, the NAB model has the worst performance (by a large difference) in comparison to MOGA and PREVIOUS models, in terms of all criteria.

Regarding test G, a new test corresponding to a non-working sunny day in summer, Model IV, a complete model, has minimum values in terms of MAE and σ. In terms of *MRE* and *MAPE*, Model IV has approximately the same performance as PREVIOUS model. In the same way as in tests D and E, the NAB model has the worst performance.

6. CONCLUSIONS

Artificial intelligence techniques are promising tools for predicting the power consumption in buildings. In this study, we applied a MOGA framework to design RBF models for this purpose and compared the performance obtained by the designed models with an already existing multilayer perceptron predictive model, proposed in [17].

Both the MOGA and PREVIOUS models aimed to predict the electric power demand in the bioclimatic CIESOL building over a prediction horizon of 1 h. MOGA models employed a sampling interval of 15 min, therefore requiring a prediction horizon of four steps, while the PREVIOUS model employed a prediction horizon of 60 steps, as data was sampled with a 1 min rate.

Four MOGA models were designed and their performance was compared with the PREVIOUS model. Model I and II used only weather variables as exogenous variables while models III, IV and PREVIOUS employed, in addition to weather information, solar cooling operation variables. According to results obtained in a battery of tests, one can say that complete models III and IV have a better performance in winter than in summer, while the performance of model I as a simple model has a compromise performance in summer and winter.

Comparing the performance of MOGA models and the PREVIOUS, despite the fact that MOGA models were trained with a small training set of 2592 samples compared to the 318340 samples used to train the PREVIOUS model, they have obtained better results, except in Test B. Moreover, as it can be seen in Table 11, the complexity of models obtained from MOGA is lower than the PREVIOUS model. According to tests D, E and G reflecting special days in winter and summer, both MOGA and PREVIOUS models have much better performance than the NAB model in terms of all criteria.

ACKNOWLEDGMENTS

Authors wish to thank the financial support provided by the Transnational Access to Research Infrastructures within the European project SFERA II (Grant Agreement N. 312.643) under the 7th Framework Program; the Project ENERPRO DPI2014-56364-C2-1-R financed by the Spanish Ministry of Economy and Competitiveness and EU-ERDF funds; the project OPTICONES financed by the IBERDROLA Spain Foundation within the framework of the 2015/16 call for Energy and Environment research grants and Portuguese Foundation for Science & Technology, through IDMEC, under LAETA, Project UID/EMS/50022/2013.

AUTHOR CONTRIBUTIONS

All authors have participated in preparing the research from the beginning to the end, such as establishing research design, method and analysis. All authors discussed and finalized the analysis results to prepare manuscript according to the progress of research.

REFERENCES

1. Perez-Lombard, L.; Ortiz, J.; Pout, C. A review on buildings energy consumption information. *Energy Build.* **2008**, *40*, 394–398.
2. Nejat, P.; Jomehzadeh, F.; Taheri, M.M.; Gohari, M.; Majid, M.Z.A. A global review of energy consumption, CO_2 emissions and policy in the residential sector (with an overview of the top ten CO_2 emitting countries). *Renew. Sustain. Energy Rev.* **2015**, *43*, 843–862.
3. Zhao, H.-X.; Magoules, F. A review on the prediction of building energy consumption. *Renew. Sustain. Energy Rev.* **2012**, *16*, 3586–3592.
4. Liang, J.; Du, R. Model-based fault detection and diagnosis of HVAC systems using support vector machine method. *Int. J. Refrig.* **2007**, *30*, 1104–1114.
5. Suganthi, L.; Samuel, A.A. Energy models for demand forecasting—A review. *Renew. Sustain. Energy Rev.* **2012**, *16*, 1223–1240.
6. Manzano-Agugliaro, F.; Montoya, F.G.; Sabio-Ortega, A.; Garcia-Cruz, A. Review of bioclimatic architecture strategies for achieving thermal comfort. *Renew. Sustain. Energy Rev.* **2015**, *49*, 736–755.
7. Gallo, C. Bioclimatic architecture. *Renew. Energy* **1994**, *5*, 1021–1027.
8. Tzikopoulos, A.F.; Karatza, M.C.; Paravantis, J.A. Modeling energy efficiency of bioclimatic buildings. *Energy Build.* **2005**, *37*, 529–544.
9. Albatici, R.; Passerini, F. Bioclimatic design of buildings considering heating requirements in Italian climatic conditions. A simplified approach. *Build. Environ.* **2011**, *46*, 1624–1631.
10. Panwar, N.L.; Kaushik, S.C.; Kothari, S. Role of renewable energy sources in environmental protection: A review. *Renew. Sustain. Energy Rev.* **2011**, *15*, 1513–1524.
11. Hua, Y.; Oliphant, M.; Hu, E.J. Development of renewable energy in Australia and China: A comparison of policies and status. *Renew. Energy* **2016**, *85*, 1044–1051.

12. Scarlat, N.; Dallemand, J.-F.; Monforti-Ferrario, F.; Banja, M.; Motola, V. Renewable energy policy framework and bioenergy contribution in the European Union—An overview from National Renewable Energy Action Plans and Progress Reports. *Renew. Sustain. Energy Rev.* **2015**, *51*, 969–985.

13. Ahmad, A.S.; Hassan, M.Y.; Abdullah, M.P.; Rahman, H.A.; Hussin, F.; Abdullah, H.; Saidur, R. A review on applications of ANN and SVM for building electrical energy consumption forecasting. *Renew. Sustain. Energy Rev.* **2014**, *33*, 102–109.

14. Fumo, N. A review on the basics of building energy estimation. *Renew. Sustain. Energy Rev.* **2014**, *31*, 53–60.

15. Kalogirou, S.A. Artificial neural networks in energy applications in buildings. *Int. J. Low Carbon Technol.* **2006**, *1*, 201–216.

16. Foucquier, A.; Robert, S.; Suard, F.; Stephan, L.; Jay, A. State of the art in building modelling and energy performances prediction: A review. *Renew. Sustain. Energy Rev.* **2013**, *23*, 272–288.

17. Mena, R.; Rodríguez, F.; Castilla, M.; Arahal, M.R. A prediction model based on neural networks for the energy consumption of a bioclimatic building. *Energy Build.* **2014**, *82*, 142–155.

18. Al-Homoud, M.S. Computer-aided building energy analysis techniques. *Build. Environ.* **2001**, *36*, 421–433.

19. Wang, S.W.; Xu, X.H. Simplified building model for transient thermal performance estimation using GA-based parameter identification. *Int. J. Therm. Sci.* **2006**, *45*, 419–432.

20. Lu, X.; Lu, T.; Kibert, C.J.; Viljanen, M. Modeling and forecasting energy consumption for heterogeneous buildings using a physical-statistical approach. *Appl. Energy* **2015**, *144*, 261–275.

21. Lomet, A.; Suard, F.; Cheze, D. Statistical modeling for real domestic hot water consumption forecasting. *Energy Procedia* **2015**, *70*, 379–387.

22. Ma, Z.; Li, H.; Sun, Q.; Wang, C.; Yan, A.; Starfelt, F. Statistical analysis of energy consumption patterns on the heat demand of buildings in district heating systems. *Energy Build.* **2014**, *85*, 464–472.

23. Fumo, N.; Biswas, M.A.R. Regression analysis for prediction of residential energy consumption. *Renew. Sustain. Energy Rev.* **2015**, *47*, 332–343.

24. Neto, A.H.; Fiorelli, F.A.S. Comparison between detailed model simulation and artificial neural network for forecasting building energy consumption. *Energy Build.* **2008**, *40*, 2169–2176.

25. Aydinalp-Koksal, M.; Ugursal, V.I. Comparison of neural network, conditional demand analysis, and engineering approaches for modeling end-use energy consumption in the residential sector. *Appl. Energy* **2008**, *85*, 271–296.

26. Ferreira, P.M.; Ruano, A.E.; Pestana, R.; Kóczy, L.T. Evolving rbf predictive models to forecast the portuguese electricity consumption. *Intell. Control Syst. Signal Process.* **2009**, *2*, 414–419.

27. Li, K.; Su, H.; Chu, J. Forecasting building energy consumption using neural networks and hybrid neuro-fuzzy system: A comparative study. *Energy Build.* **2011**, *43*, 2893–2899.

28. Karatasou, S.; Santamouris, M.; Geros, V. Modeling and predicting building's energy use with artificial neural networks: Methods and results. *Energy Build.* **2006**, *38*, 949–958.

29. Kaytez, F.; Taplamacioglu, M.C.; Cam, E.; Hardalac, F. Forecasting electricity consumption: A comparison of regression analysis, neural networks and least squares support vector machines. *Int. J. Electr. Power Energy Syst.* **2015**, *67*, 431–438.

30. Dong, B.; Cao, C.; Lee, S.E. Applying support vector machines to predict building energy consumption in tropical region. *Energy Build.* **2005**, *37*, 545–553.

31. Jung, H.C.; Kim, J.S.; Heo, H. Prediction of building energy consumption using an improved real coded genetic algorithm based least squares support vector machine approach. *Energy Build.* **2015**, *90*, 76–84.

32. Suganthi, L.; Iniyan, S.; Samuel, A.A. Applications of fuzzy logic in renewable energy systems—A review. *Renew. Sustain. Energy Rev.* **2015**, *48*, 585–607.

33. Ciabattoni, L.; Grisostomi, M.; Ippoliti, G.; Longhi, S. Fuzzy logic home energy consumption modeling for residential photovoltaic plant sizing in the new Italian scenario. *Energy* **2014**, *74*, 359–367.

34. Li, K.; Su, H. Forecasting building energy consumption with hybrid genetic algorithm-hierarchical adaptive network-based fuzzy inference system. *Energy Build.* **2010**, *42*, 2070–2076.

35. Hamzacebi, C.; Es, H.A. Forecasting the annual electricity consumption of Turkey using an optimized grey model. *Energy* **2014**, *70*, 165–171.

36. Lee, Y.-S.; Tong, L.-I. Forecasting energy consumption using a grey model improved by incorporating genetic programming. *Energy Convers. Manag.* **2011**, *52*, 147–152.

37. Guo, J.J.; Wu, J.Y.; Wang, R.Z. A new approach to energy consumption prediction of domestic heat pump water heater based on grey system theory. *Energy Build.* **2011**, *43*, 1273–1279.

38. Dagnely, P.; Ruette, T.; Tourwe, T. Predicting hourly energy consumption. Can you beat an autoregressive model? In Proceeding of the 24th Annual Machine Learning Conference of Belgium and the Netherlands, Benelearn, Delft, The Netherlands, 19 June 2015.

39. *Estación Experimental "Las Palmerillas". Datos Meteorológicos. Campañas Agrícolas 76/77–94/95*; Caja Rural de Almería: Almería, Spain, 1997.

40. Castilla, M.Á.; Álvarez, J.D. *Francisco de Asis Berenguel, Manuel. Comfort control in Buildings*, 1st ed.; Springer London: London, UK, 2014.

41. Heras, M.R.; Jimenez, M.J.; San Isidro, M.J.; Zarzalejo, L.F.; Perez, M. Energetic analysis of a passive solar design, incorporated in a courtyard after refurbishment, using an innovative cover component based in a sawtooth roof concept. *Sol. Energy* **2005**, *78*, 85–96.

42. Ferreira, P.M.; Ruano, A.E. Evolutionary multiobjective neural network models identification: Evolving task-optimised models. *New Adv. Intell. Signal Process.* **2011**, *372*, 21–53. [Google Scholar]

43. Ruano, A.E.; Ferreira, P.M.; Fonseca, C.M. An overview of nonlinear identification and control with neural networks. In *Intelligent Control Systems Using Computational Intelligence Techniques*; Ruano, A.E., Ed.; Institution of Electrical Engineers: London, UK, 2005; pp. 37–87. [Google Scholar]

44. Marquardt, D.W. An algorithm for least-squares estimation of nonlinear parameters. *J. Soc. Ind. Appl. Math.* **1963**, *11*, 431–441. [Google Scholar] [CrossRef]

45. Hornik, K.; Stinchcombe, M.; White, H. Multilayer feedforward networks are universal approximators. *Neural Netw.* **1989**, *2*, 359–366. [Google Scholar] [CrossRef]

46. Fraser, A.M.; Swinney, H.L. Independent coordinates for strange attractors from mutual information. *Phys. Rev. A* **1986**, *33*, 1134–1140. [Google Scholar] [CrossRef] [PubMed]

47. Kennel, M.B.; Brown, R.; Abarbanel, H.D.I. Determining embedding dimension for phase-space reconstruction using a geometrical construction. *Phys. Rev. A* **1992**, *45*, 3403–3411. [Google Scholar] [CrossRef] [PubMed]

48. Zhou, A.; Qu, B.-Y.; Li, H.; Zhao, S.-Z.; Suganthan, P.N.; Zhang, Q. Multiobjective evolutionary algorithms: A survey of the state of the art. *Swarm Evolut. Comput.* **2011**, *1*, 32–49. [Google Scholar] [CrossRef]

49. Zitzler, E.; Deb, K.; Thiele, L. Comparison of multiobjective evolutionary algorithms: Empirical results. *Evolut. Comput.* **2000**, *8*, 173–195. [Google Scholar] [CrossRef] [PubMed]

50. Fonseca, C.M.; Fleming, P.J. An overview of evolutionary algorithms in multiobjective optimization. *Evolut. Comput.* **1995**, *3*, 1–16. [Google Scholar] [CrossRef]

51. Fonseca, C.M.; Fleming, P.J. Multiobjective optimization and multiple constraint handling with evolutionary algorithms. I. A unified formulation. *IEEE Trans. Syst. Man Cybern. Part A Syst. Hum.* **1998**, *28*, 26–37. [Google Scholar] [CrossRef]

52. Ruano, A.E.B.; Jones, D.I.; Fleming, P.J. A new formulation of the learning problem for a neural network controller. In Proceeding of the 30th IEEE Conference on Decision and Control, Brighton, UK, 11–13 December 1991; pp. 865–866.

53. Ferreira, P.M.; Faria, E.A.; Ruano, A.E. Neural network models in greenhouse air temperature prediction.*Neurocomputing* **2002**, *43*, 51–75.

54. Ferreira, P.M.; Ruano, A.E.; Ieee, I. Exploiting the separability of linear and nonlinear parameters in radial basis function networks. In Proceeding of the IEEE 2000 Adaptive Systems for Signal Processing, Communications, and Control Symposium, Lake Louise, AB, Canada, 1–4 October 2000; pp. 321–326.

55. Haykin, S. *Neural Networks: A Comprehensive Foundation*, 2nd ed.; Prentice Hall: Upper Saddle River, NJ, USA, 1999.

56. Ruano, A.; Khosravani, H.R.; Ferreira, P.M. A randomized approximation convex hull algorithm for high dimensions. *IFAC PapersOnLine* **2015**, *48*, 123–128.

CHAPTER 9

Model Predictive Control-Based Fast Charging for Vehicular Batteries

Jingyu Yan [1,2], Guoqing Xu [1,2,3,*], Huihuan Qian [1,2], Yangsheng Xu [1,2] and Zhibin Song [1]

[1]Shenzhen Institutes of Advanced Technology, The Chinese Academy of Science, Shenzhen 518055, China
[2]Department of Mechanical and Automation Engineering, The Chinese University of Hong Kong, Shatin, Hong Kong, China
[3]Department of Electrical Engineering, Tongji University, Shanghai 200092, China

ABSTRACT

Battery fast charging is one of the most significant and difficult techniques affecting the commercialization of electric vehicles (EVs). In this paper, we propose a fast charge framework based on model predictive control, with the aim of simultaneously reducing the charge duration, which represents the out-of-service time of vehicles, and the increase in temperature, which represents safety and energy efficiency during the charge process. The RC model is employed to predict the future State of Charge (SOC). A single mode lumped-parameter thermal model and a neural network trained by real experimental data are also applied to predict the future temperature in simulations and experiments respectively. A genetic algorithm is then applied to find the best charge sequence under a specified fitness function, which consists of two objectives: minimizing the charging duration and minimizing the increase in temperature. Both simulation and experiment demonstrate that the Pareto front of the proposed method dominates that of the most popular constant current constant voltage (CCCV) charge method

Keywords: battery fast charging; model predictive control; state of charge; genetic algorithm; electric vehicles

1. INTRODUCTION

It is necessary to refuel all vehicles after a period of travel. For conventional fuel-driven vehicles, the fuel can be replenished quickly and safely at filling stations, with a minimal out-of-service time and without energy loss. However, the replenishment of energy in electric vehicles requires charging the battery pack. In general applications, such as MP3 players, shavers, cell phones, the charging process may last several hours, leading to a long out-of-service time. The charging method is also crucial; an unsuitable method leads to high increase in temperature in a battery and even the risk of fire and explosion.

Therefore, charging control is a significant issue in battery management systems, with the aim of feeding external energy into batteries in a fast, safe, and efficient way. Fast charging helps to reduce out-of-service time and promote the commercialization of EVS. Safe charging not only assures the safety of users by preventing battery burning and explosions during the charging process, but also prolongs the battery life by preventing overcharging and overheating damage. Efficient charging can convert as much electrical energy as possible from a charger to electrochemical energy stored in a battery so as to enhance energy efficiency.

1.1. Literature Review

Charging methods have been studied since the invention of rechargeable batteries. The earliest and the most common charging method is named the constant trickle current (CTC) [1] method, which has a simple circuit structure and very low cost. Thus, the constant trickle current method has been adopted in most electronic products for many years. Because this method provides a very small current, the charging time is extremely long so that it usually works overnight. An easy way to reduce the charging time is to increase the constant current (CC). Charging with 1 C current can fill an empty battery in one hour. However, one drawback in the CC charging method is that it requires a very precise state of charge (SOC) estimator which has the ability to determine whether the battery is fully charged and whether to stop the charging process. Unfortunately, the precise SOC estimator is not easy to implement. Another disadvantage is that CC charging cannot avoid over-voltage to the battery. A battery can be simply modeled with an open circuit voltage (OCV) source and a series-connected resistor. The voltage on the resistor will cause the terminal voltage to be always higher than the OCV during CC charging, finally leading to over-voltage when the OCV approaches its full value. Taking a Li-ion battery as an example, the over-voltage during charging will degrade the crystallographic structure of the cathode and cause oxidative decomposition of the electrolytic solvents [2].

The constant current constant voltage (CCCV) method has been proposed to overcome the disadvantages of CC charging. A CC period is applied until the charging voltage reaches a predetermined value, and then the charging process goes into the CV period. In the CV period, the charging voltage is fixed to a cutoff value; therefore, the charging current will be automatically reduced along with the increase of SOC. Generally, the CV period requires a long charging period [3]. In order to enhance charging performance of CCCV, various combinations of CC and CV periods have been proposed in recent years, such as CCCVCV [4], $(CCCV)^2$ [5], and so forth. A smooth control circuit (SCC) is proposed to ensure the stable transition from the constant-current (CC) to the constant-voltage (CV) stage [6]. A further review of CCCV family is given in [7].

In recent years, since microprocessor control units (MCUs) and even computers have been applied to monitor and control the charging process at charging station for EVs, so some intelligent methods can be put into practice. One kind of intelligent charger controls the charging process by detecting the tuning points. The tuning points are selected as the threshold points, stationary points, and inflection points of battery voltage, temperature, and lapsed time [8,9,10]. These intelligent methods only change the charging behavior at tuning points and the periods between tuning points are still under open-loop control. Fuzzy control has been also applied to solve the charging control problem [3,11]. Optimal control based on power loss model was adopted to enhance the charging efficiency [12]. In [13,14,15], neural networks and genetic algorithms are also introduced to design the membership functions and inference rules of the fuzzy controller. Optimization methods, such as the ant colony algorithm and evolutionary algorithm, have also been introduced to optimize a best multi-stage CC charging profile [16,17]. The optimization does not aim to regulate the real-time charging current according to instant system measurements, but rather to find a best charging profile covering the whole charging process. Therefore, its performance is seriously dependent on the accuracy of the model used in optimization. Suffering from system noise and the time-variance properties of batteries, the robustness of these methods is weak. In [18], to increase the charging speed, a grey prediction technique is utilized to develop a grey-predicted control system. In that study, the control system does not take increase in temperature into consideration and only applies a one-step prediction. The small predictive horizon weakens the prediction of future system behavior.

1.2. Overview of Proposed Charging Controller

Modeling of batteries has been comprehensively studied in recent years, and some are demonstrated to be accurate and efficient enough to model the behaviors of batteries [19,20,21,22,23]. These models make it possible to apply model predictive control (MPC) to manage the battery charging process.

MPC is an advanced control method and has been widely applied in many fields [24,25]. The working principle of MPC is to apply system models to predict system responses to possible future control sequences and to find the best future control sequence by optimizing the user-defined objective function. Only the first element of the control sequence is applied to the controlled system at each time. The system states are then sampled again and the calculations are repeated at the next control time. The prediction horizon is constantly shifted forward. Because the receding horizon strategy updates predictions based on the instant measurement of system inputs and outputs at each control time, its robustness has been demonstrated to be strong [26]. MPC may be categorized as either linear MPC or nonlinear MPC, according to the linearity of the system models and constraints. For linear MPC, the best control sequence can be analytically obtained by solving Diophantine equations. However, for nonlinear MPC, the best control sequence usually is calculated numerically under an optimization framework [27].

Available battery models establish the basis for applying a MPC framework to solve the charge control problem. Three advantages of MPC explain the rationale. Firstly, it has the ability to predict future battery states under a possible future charging sequence and hence evaluate charging performance over a comparatively long period. Secondly, it calculates the best charging sequence using an optimization method, which can solve multi-objective problems. Thirdly, it utilizes a receding horizon strategy and only applies the first element of the best charging sequence at each control moment. The prediction error caused by model inaccuracy will not accumulate because instantaneous measurements correctly update each initial prediction value. As shown in Figure 1, the fast charging control framework proposed in this work consists of the following components:

1. A SOC predictor, predicting the SOC of battery when fed by a sequence of future charging current;
2. A temperature predictor, predicting the future battery temperature under a sequence of charging current;
3. A fitness evaluator, evaluating the performance of the sequence of charging current;
4. An optimizer, finding the best sequence of charging current using genetic algorithm (GA).

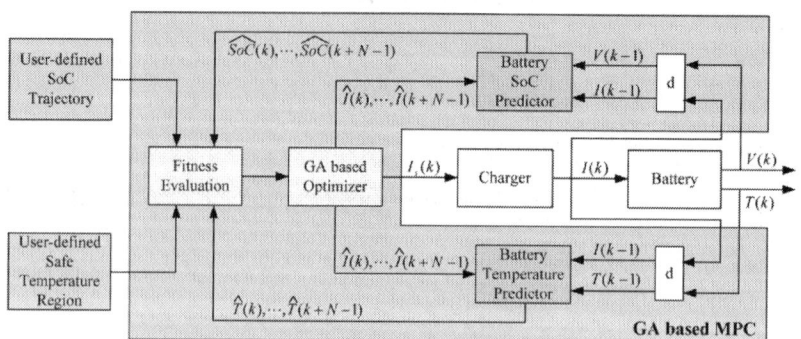

Figure 1. Fast charging control framework based on model predictive control.

2. PREDICTIVE MODELS

To utilize predictive control, we apply battery models to predict future battery states under a sequence of future charging current. The battery states of interest in this study are SOC and temperature. The former indicates the charging duration and whether overcharge damage is possible, while the latter represents the safety and energy efficiency during the charging process. Charging control is necessary to: (1) minimize charging duration; (2) minimize temperature increases; and (3) restrict temperature to a safe range.

2.1. RC Model for SOC Prediction

Figure 2 describes the RC equivalent circuit model [28], which is applied in this paper. This model consists of a bulk capacitor Cb which simulates energy storage and a surface capacitorCs which represents the dynamic property of the battery. Output resistance R0, surface resistance Rs and bulk resistance Rb are utilized to model the internal resistance of the battery.

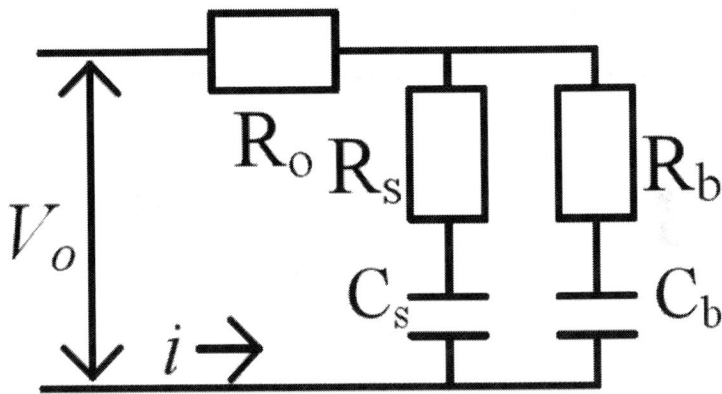

Figure 2. The battery RC Model.

In this paper, we denote k and Ts as the time index and sampling time, Vb(k) and Vs(k) as the voltages of bulk capacitor and surface capacitor respectively, I(k) as the charging current,V0(k) as the terminal voltage. The voltage of bulk and surface capacitor are selected as the state vectors. Charging current is selected as the system input and terminal voltage is selected as the system output. System discrete time-variant state space equations are expressed as follows:

$$x(k+1) = A(k)x(k) + B(k)u(k) \qquad (1)$$

$$y(k) = C(k)x(k) + D(k)u(k) \tag{2}$$

where:

$$x(k) = [V_b(k), V_s(k)]^T \tag{3}$$

$$u(k) = I(k) \tag{4}$$

$$y(k) = V_0(k) \tag{5}$$

$$A(k) = \begin{bmatrix} 1 - \dfrac{T_s}{C_b(k)[(R_b(k) + R_s(k)]} & \dfrac{T_s}{C_b(k)[R_b(k) + R_s(k)]} \\ \dfrac{T_s}{C_s(k)[R_b(k) + R_s(k)]} & 1 - \dfrac{T_s}{C_s(k)[R_b(k) + R_s(k)]} \end{bmatrix} \tag{6}$$

$$B(k) = \begin{bmatrix} \dfrac{-R_s(k)T_s}{C_b(k)[R_b(k) + R_s(k)]} \\ \dfrac{-R_b(k)T_s}{C_s(k)[(R_b(k) + R_s(k)]} \end{bmatrix} \tag{7}$$

$$C = \begin{bmatrix} \dfrac{R_s(k)}{R_b(k) + R_s(k)} & \dfrac{R_b(k)}{R_b(k) + R_s(k)} \end{bmatrix} \tag{8}$$

$$D = -R_0(k) - \dfrac{R_b(k)R_s(k)}{R_b(k) + R_s(k)} \tag{9}$$

SOC in the RC model is estimated using the voltages of the two capacitors based on the relationship between SOC and the open-circuit voltage (OCV). Since Cb represents the bulk energy in the battery, it contributes the majority of the battery SOC, as expressed as follows:

$$\overline{SOC} = \dfrac{1}{21}[20\overline{SOC_{C_b}}(k) + \overline{SOC_{C_s}}(k)] \tag{10}$$

where:

$$\overline{SOC_{C_b}}(k) = F_{OCV-SOC}(V_b(k)) \tag{11}$$

$$\overline{SOC_{C_s}}(k) = F_{OCV-SOC}(V_s(k)) \tag{12}$$

$F_{OCV-SOC}(\cdot)$ is the function mapping OCV to SOC. It is usually predetermined by the manufacturer's datasheet or experimental data.

2.2. Thermal Model in Simulation

Reference [28] proposed a simple single-node lumped-parameter thermal modal for batteries. It models the thermal process in three stages. In the first stage, the Joule effect generates heat in the battery. In the second, the battery's heat is conducted and convected to the surrounding air. Finally, the surrounding air exchanges heat with the ambient. In the RC model, the heat generation is expressed by:

$$Q_g(k) = T_s[I_0^2(k)R_0(k) + I_s^2(k)R_s(k) + I_b^2(k)R_b(k)] \tag{13}$$

where:

$$I_0(k) = I(k) \tag{14}$$

$$I_b(k) = \frac{V_b(k) - V_s(k) + R_s(k)I(k)}{R_s(k) + R_b(k)} \tag{15}$$

$$I_s(k) = I_b(k) - I(k) \tag{16}$$

Meanwhile, the heat passing from the battery to the surrounding air is expressed by:

$$Q_p(k) = \frac{T(k-1) - T_{air}(k-1)}{R_{eff}} \tag{17}$$

where T_{air} represents the effective temperature of surrounding air and Reff stands for the effective thermal resistance. Based on the assumption that 50% of the heat from the battery is spent to warm the air [28], Tair is expressed by:

$$T_{air}(k-1) = T_{amb} - \frac{0.5 \times Q_p(k-1)}{\dot{m}_{air}C_{air}} \tag{18}$$

where Tamb is the ambient temperature, Qp(k−1) is the passing heat in previous step, m˙air is the airflow rate, and Cair is its heat capacity.

Strictly speaking, the value of Reff depends on the thermal control method. For example, it becomes smaller if the cooling fans are open. However, in this study, we assume the charging process is conducted in an environment with only natural convection. In this case, Reff is fixed as a constant with the value calculated by:

$$R_{eff} = \frac{1}{hA} + \frac{t}{kA} \tag{19}$$

where h is the heat transfer coefficient in the natural environment, A is the total module surface area exposed to the air, t is the thickness of the module case, and k is the thermal conductivity of the module case material.

We denote Qg(k) as the heat generation which heats the battery and Qp(k) as the heat loss which cools the battery by heating the surrounding air. Therefore, the battery temperature T(k) can be calculated by:

$$T(k) = T(k-1) + \frac{Q_g(k) - Q_p(k)}{m_{bat} C_{bat}} \tag{20}$$

where mbat is the mass of the battery, Cbat is heat capacity of the battery.

2.3. Thermal Model in Experiment

The simulation model requires many physical parameters of the battery and working environment. However, in practice, these parameters usually cannot be obtained accurately. Meanwhile, the theoretical model only considers the ideal situation so that it may not be suitable for the complex and nonlinear electrochemical process inside the battery. Furthermore, heat generation is only caused by heating of the resistors, which does not fully represent the actual situation. In reality, heat generation during charge is also affected by the charging acceptance rate. In [5], it pointed out that "close-to-fully discharged batteries can be recharged with very high currents for a short period of time" and [29] concluded that for A123, a kind of LiFePO4, "at a low state of charge, nearly all the charging current is absorbed by the chemical reaction. Above 80% of SOC, more and more energy goes into heat".

Figure 3 shows the temperature curves of our experimental LiFePO4 cells during charge processes with different charging rates. The increase in temperature agrees with the two results reported in the literature above.

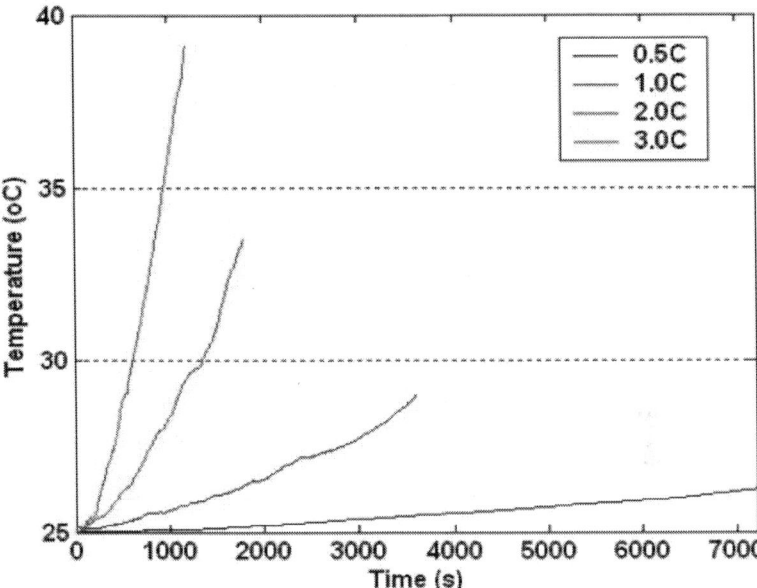

Figure 3. Increase in temperature under different charging rates.

The increase in temperature obviously also has a close relationship with the charging rate. High charging rates will cause high increase in temperature. Therefore, we select the SOC, the charging rate and the current temperature to predict the future temperature based on a neural network (NN).

Figure 4 shows the structure of the NN. The training set consists of the real data recorded in the charging process, as shown in Figure 3. The input vector contains three parameters of interest: SOC(k), T(k) and I(k), and the output or the predictive result is T(k+1). T is directly measured by a temperature sensor while I and SOC is measured and estimated respectively by the programmable charger IT6154, measurement of current of which is accurate enough to calculate SOC using columbic counting method. Since the whole charging process is recorded, the input vector at time k and the output at time k+1 are known to train the neural network.

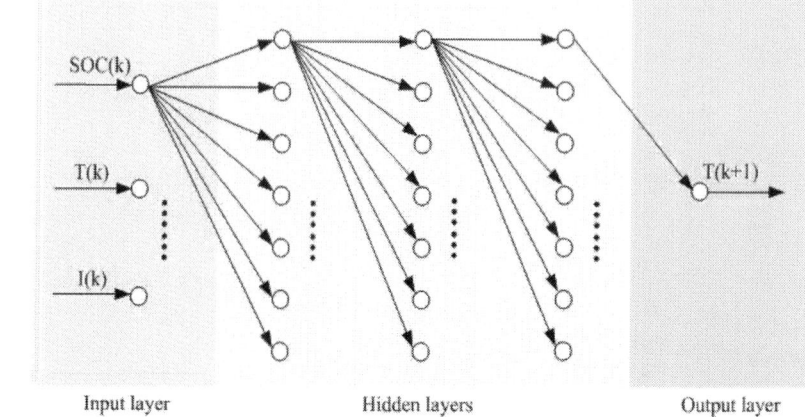

Figure 4. Neural network for predicting battery temperature.

The testing set also consists of the real data recorded in charging process, including the batteries used in training set and those only used in testing set. The predictive error is defined as the expression:

$$\left| \frac{\hat{T}(k+1) - T(k+1)}{T(k+1)} \right| \times 100\%$$

where T^(k+1) is the predictive result by the neural network and T(k+1) is the measured result during the charging process. The training method is back propagation (BP) learning. When the time step is set to 30s, the average prediction error is 3.36% and the maximum is 5.43%.

2.4. Model Based Prediction

To predict battery states under a given sequence of future charging current, two problems should be solved: (1) how to initialize the prediction at time k; and (2) how to realize the multi-step prediction based on the one-step prediction model.

In order to initialize the prediction at time k, the RC model must know the last system statex(k−1), the last system input u(k−1), and the model parameters at time k. Among them, u(k−1) and y(k−1) can be obtained from the current sensor and voltage measurement, respectively, while x(k−1) can be obtained by closed-loop estimators such as the extended Kalman filter [30], sigma-point Kalman filter [31,32], H∞ filter [33], and so on. The closed-loop filters are able to eliminate accumulated errors and the estimation result will gradually converge to the real value, or to a limited error range. The model parameters are stored in a look-up table indexed by SOC and temperature. The table is constructed using the off-line testing data. The slow changes in SOC and temperature give the reason for updating the model parameters at time k according SOC(k−1) and T(k−1), especially when Ts is comparatively small. SOC(k−1) is calculated according to Equation (10), and T(k−1) is measured directly from a temperature sensor fixed on the surface of the battery.

Given the value of I(k), the above preparation allows a one-step prediction of SOC(k) andT(k). However, a single-step prediction is usually insufficient to predict system behavior over a long process. Generally speaking, the prediction horizon in MPC is more than one step. Therefore, a multi-step predictor is necessary to predict system states exerted by a sequence of future charging current. A simple way to realize the multi-step predictor is an iterative prediction, which uses the predictive future system states at time k+j as the initial state for the next time k+j+1.

In summary, at control time k, the SOC model and temperature model can iteratively map a sequence of future charging current [I(k),I(k+1),...,I(k+p−1)] to future battery states[SOC(k),SOC(k+1),...,SOC(k+p−1)] and [T(k),T(k+1),...,T(k+p−1)]. The multi-step prediction uses an open-loop manner in which each step suffers from the prediction error in the previous step and finally reverts to the errors in the initial values at time k−1. Fortunately, these initial values are estimated in a closed-loop manner so that initial errors are comparatively small. In reality, any battery management system requires such an estimator to obtain the real-time SOC and measure the battery temperature. The results can be used as the initial values.

3. FORMULATION UNDER MPC FRAMEWORK

Given a sequence of future inputs, the length of which is denoted as the prediction horizon P, the future system states can be predicted based on the dynamic system model. The future system behavior under the sequence of inputs can then be evaluated based on a performance index. At each control time k, the basic idea of MPC is to find an optimal sequence of inputs $[\hat{u}(k),\hat{u}(k+1),...,\hat{u}(k+p-1)]$, which optimizes the performance index, and then apply the first element of the input sequence u^(k) to the system as the current control variable. In the charging control problem, MPC optimizes a sequence of future charging current [I(k),I(k+1),...,I(k+p−1)], which has the best performance index based on the predicted battery $[\overline{SOC}(k),\overline{SOC}(k+p-1),...,\overline{SOC}(k+p-1)]$ and $[\hat{T}(k),\hat{T}(k+1),...,\hat{T}(k+p-1)]$.

3.1. Performance Indexes

The performance index reflects control objectives. The first objective is to minimize charging duration. However, in a limited prediction horizon, it is impossible to directly predict the whole charging process, so we turn to the tracking of a user-defined SOC trajectory instead. A fast-rising SOC trajectory requires a fast charging speed while a flat one requires a slow speed.

The expected SOC trajectory can directly copy from any real charging trajectory controlled by any charging scheme. In addition, the expected trajectory can be set as a real trajectory with revisions based on some special considerations. To track the expected trajectory, the performance index J1 is expressed by:

$$J_1 = \left| \overline{SOC}\left(k+p-1\right) - SOC^*\left(k+p-1\right) \right| \tag{21}$$

where SOC^* is the user-defined, expected SOC trajectory. Because SOC is required to achieve the expected point in the final step, evaluation of the SOC tracking performance is based on the final prediction state only. How this is achieved is not very important from the point of view of the charging process.

The second objective of charging control is to minimize the increase in temperature, which partially reflects energy efficiency and system safety. Assuming that the expected trajectory is copied from a CCCV charging process, without consideration of temperature, the best charging current sequence is the same sequence applied in the CCCV process. However, if we take increase in temperature into consideration to evaluate charging performance, as expressed in J2, the MPC will try to track the expected SOC curve using a process in which increase in temperature is minimized:

$$J_2 = \max\{\hat{T}(k+j) - T(k-1) \mid j = 0,1,...,p-1\} \tag{22}$$

3.2. Constraints

Because a charging sequence may cause damage to the battery or lead to dangerous events, it is necessary to design some constraints to protect the battery during the charging process. The first constraint is that the SOC must not exceed 100%, to avoid over-charge damage. In practice, to reserve some tolerance, 98% SOC is treated as the full state. The second constraint is that temperature must be kept in a user-defined range to avoid overheating caused by overcharge or by a large charging current which exceeds the instant charging acceptance level. The two constraints are expressed by:

$$C_1 : \overline{SOC}(k+j) \le 98\% \tag{23}$$

and:

$$C_2 : \hat{T}(k+j) \le \overline{T}(k+j) \tag{24}$$

where $j=0,1,...,p-1$ and $\bar{T}$ is the user-defined safe range, which is designed either as a constant, indicating the highest temperature during the whole charging process, or as a time indexed function specifying the temperature limitation along with charging duration.

4. OPTIMIZATION USING GENETIC ALGORITHM

The MPC charging control problem is formulized to minimize the performance indexes J1 and J2, subject to the constraints C1 and C2. Essentially, the control problem is transformed into a constrained multi-objective optimization problem.

Generally, a multi-objective optimization problem can either be solved by multi-objective optimizers directly, or by transforming it into a single-objective problem which can be solved by single-objective optimizers [34]. In this paper, we apply the latter method because SOC tracking is more important than temperature management in the charging process, so the two objectives can be merged into one index by summing with different weights. To solve the optimization problem, a genetic algorithm (GA) is applied because of its strong global searching ability without the requirement of derivative information of objective functions. Since the two constraints should be strictly satisfied in the charging process to assure the safety and health of the battery, solutions that fail to satisfy anyone of the constraints will be assigned the worst fitness.

From what discussed above, the fitness function to be minimized in GA is expressed by:

$$F = \begin{cases} \omega_1 J_1 + \omega_2 J_2 & \left(C_1 \text{ and } C_2 \text{ are held}\right) \\ +\infty & \left(\text{otherwise}\right) \end{cases} \tag{25}$$

where ω_1 and ω_2 are the weights of J_1 and J_2.

The minimization problem is solved by a standard GA, the scheme of which is illustrated inFigure 5 and briefly described in the following steps.

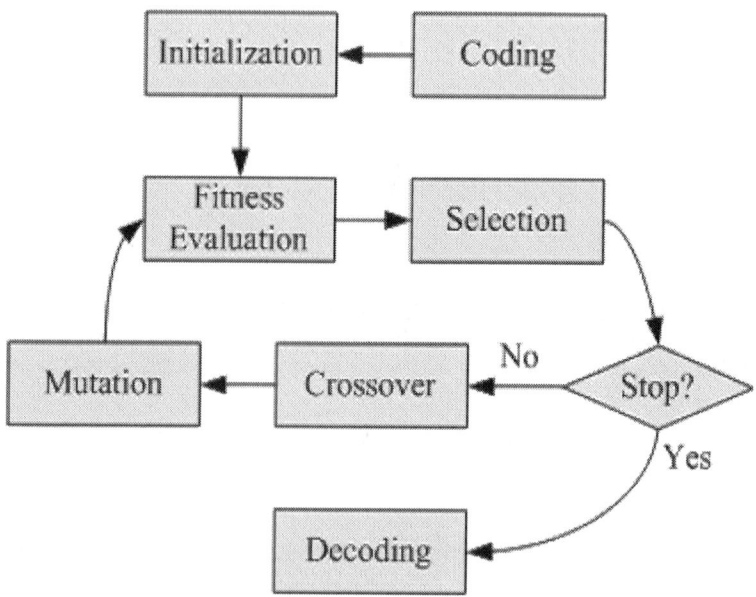

Figure 5. Scheme of the standard GA.

1. Coding. The standard GA generally codes a candidate solution as a string of characters which are usually binary digits, referred to as a chromosome. The candidate solution is termed an individual. Accordingly, the set made up of a number of individuals is termed a population. In this paper, we apply a real-value coding method, which codes a candidate solution as a set of floating decision variables. The real-value coding method is proven to have superior performance to the binary-coded method in control optimization problems [35].

2. Initialization. The standard GA starts with an initial population. Usually, individuals in the initial population are produced randomly. In MPC, the initialization process is executed at each control time to start the GA. Since the best control sequence optimized at time k contains good candidates from $k+1$ to $k+p-1$, one initial individual is specially designed by shifting it one time step and filling the last charging current with the same value as $I^{\wedge}(k+p-1)$, as shown in Figure 6. This individual introduces historical best charging sequence into the current optimization process, thus it is helpful to improve optimization performance to be at least very similar with the previous optimized performance.

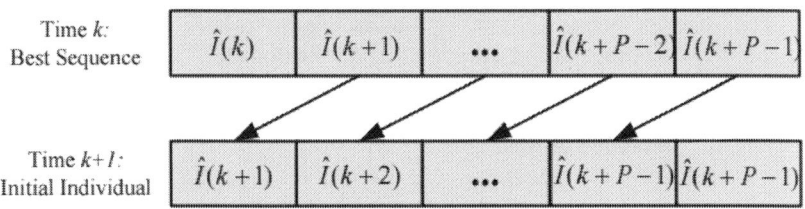

Time k:
Best Sequence

| $\hat{I}(k)$ | $\hat{I}(k+1)$ | ... | $\hat{I}(k+P-2)$ | $\hat{I}(k+P-1)$ |

Time $k+1$:
Initial Individual

| $\hat{I}(k+1)$ | $\hat{I}(k+2)$ | ... | $\hat{I}(k+P-1)$ | $\hat{I}(k+P-1)$ |

Figure 6. Initialization of one special individual by introducing the best control sequence optimized in previous step into the present step.

3. Fitness evaluation. We evaluate the fitness of each individual in each generation according to the Equation (25). The smaller the fitness, the better the individual. However, to facilitate the following selection step, the raw fitness is usually scaled to assign suitable selection pressure to each individual. In this work, the scale function is expressed as follows:

$$F_{scale} = \frac{1}{1+F^r}$$
(26)

where r is power of raw fitness. A large r will quickly increase the selection pressure to a worse individual, accelerate the convergence speed, and increase the risk of premature especially for multi-peak landscape, and *vice versa*.

4. Selection. Individuals are selected from the previous generation to the current generation based on the scaled fitness Fscale, following the survival of the fittest rule. Many selection methods have been developed to avoid genetic drift and premature phenomena. In this work, the roulette wheel selection method is adopted [36]. The elitism strategy is also applied in selection to assure that the best solution will never be lost.

5. Crossover. In the crossover step, the standard GA exchanges information between two parent individuals and produces two child individuals. In this work, the arithmetical crossover method is used. Given two parents x1 and x2, the children y1 and y2 are produced by linear combinations of parents with a random coefficient λ:

i. $$y_1 = \lambda x_1 + (1-\lambda)x_2$$
(27)

ii. $$y_2 = \lambda x_2 + (1-\lambda)x_1$$
(28)

6. Mutation. After the crossover step, a subset of individuals is selected with a mutation probability of pm. To explore the search space, we use Gaussian mutation, which adds a random value from a Gaussian distribution with variance σ to each item of the selected individual.

7. Termination. Many terminating conditions have been proposed to stop the iteration process. For example, when the distances among individuals are smaller than a predetermined value, an individual satisfies a minimum criterion, or the maximum number of generations is reached. The last method is applied here.

5. PERFORMANCE DEMONSTRATION

5.1. Settings

In order to evaluate the proposed MPC based charging control strategy, simulations are conducted based on a well established "7 Ah Saft Lithium-Ion battery" provided in Advisor [21]. RC model parameters and OCV all are time-variant variables depending on SOC and temperature. The curves of OCV and R0 are shown in Figure 7 as examples. Interpolating method is applied to create the continuous values space. The constant parameters in this simulation environment are given in Table1.

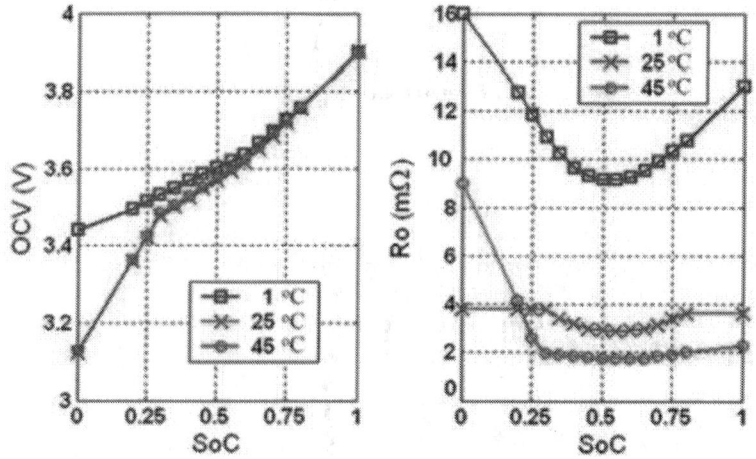

Figure 7. Battery time-variant properties. Taking OCV and Ro as examples. Data source: Advisor.

In this work, we select the control period Ts to be 30 seconds using a trial and error method according to experimental results. A shorter period requires the charger to change the charging current more frequently. It will increase working load to the charger and lead to more energy loss caused by switch circuits. The longer period will decrease the frequency to tune charging current in the charging process so that the performance of MPC will be limited. In addition, the long control period implies that the prediction based on the model is over a long time which essentially requires a more accurate battery model. Finally, the 30 seconds control period is long enough to allow GA to finish the optimization process.

Table 1. Parameters setting in the simulation and experiment.

	Symbol	Description	Value	Unit
Battery (simulation)	C	Battery nominal capacity	7	Ah
	m_{bat}	Battery mass	0.37824	kg
	C_{bat}	Battery heat capacity	795	J/kgK
	R_{eff}	Effective thermal resistance	7.8146	K/W
	T_{amb}	Ambient temperature	20	°C
	$\dot{m}_{air}$	Airflow rate	5.8333	g/s
	C_{air}	Air heat capacity	1009	J/kgK
MPC	T_s	Control period	30	s
	p	Prediction horizon	5	--
	ω_1	Weight of SOC tracking J_1	100	--
	ω_2	Weight of termperature rising J_2	1	--
GA	$MaxGen$	Maximum generation number	30	--
	$Popsize$	Population size	50	--
	r	Power of raw fitness in scaling	2	--
	R_λ	The range of crossover coefficient	[0.1, 0.9]	--
	p_m	Mutation probability	0.2	--
	σ	Variance of Gaussian mutation	1	--

Remark: The typical value of J_1 is around 0.08 while that of J_2 is around 2. Therefore, the real weight ratio of J_1 to J_2 is around 4:1.

5.2. Evaluation Method

Fast charging speed and low heat generation are both objectives in charging control. However, the two goals conflict with each other. Fast charging essentially requires large current and hence leads to high heat generation, and vice versa. From the viewpoint of multi-objective optimization, the conflicting objectives are usually evaluated by Pareto curve. For charging control, the x-axis is set as the charging duration and the y-axis as the final temperature increase T(end)−T(0). As shown in Figure 8, the curve with the circle marks represents the Pareto front of the CCCV family, where the CC period applies 1.5 C to 6 C current for fast charging. The CCCV Pareto front splits the objective space into two sections. Any charging result located in the upper right section is worse than the CCCV family for both objectives, while any result in the bottom left section is better than CCCV for both objectives. Therefore, we evaluate the performance of the charging controller according to the location of results.

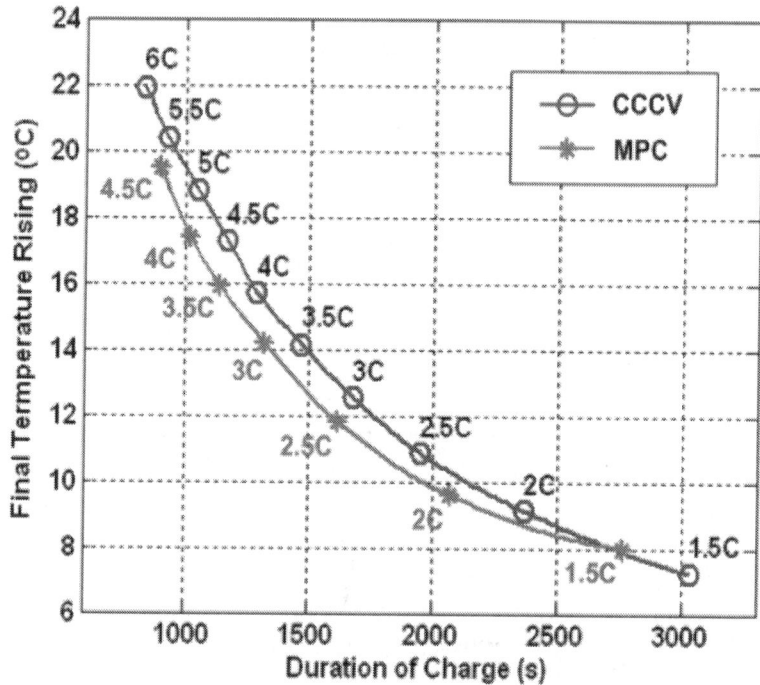

Figure 8. Pareto fronts of CCCV and MPC charging methods in simulation. The expected trajectories of MPC are modified from CCCV by multiplying 1.05.

5.3. Simulation Results

To facilitate comparison with the conventional CCCV family, we set the expected SOC trajectory by multiplying by 1.05. The new trajectory is intended to accelerate the charging speed. Since the working temperature of Li-ion batteries, especially for vehicular batteries, is from 20 °C to 40 °C, the ambient temperature is set as 20 °C and the safe range of temperature is set below 40 °C. The search space of the charging current is fixed from 0 C to 6 C.

The MPC is applied to track the modified SOC trajectories from 1.5 C to 4.5 C. A faster charging speed cannot keep the rising temperature within a safe range. The Pareto front of the MPC is illustrated in Figure 8 by the curve marked with stars. The results for the MPC clearly dominate those of the CCCV family. Only the result of the trajectory revised from 1.5 C has a similar performance to CCCV. The reason is that a 1.5 C current is comparatively so small that it limits the applicable current sequence to a small range. If the fixed search range can be accordingly reduced, e.g., from 0 C to 3 C, the result will be improved.

As an example, the charging processes of CCCV and MPC are compared in the case of 3 C, shown in Figure 9. At the beginning of charge, the internal resistance is large when the SOC is very low. In this process, the rising temperature dominates the fitness function. The optimized charging profile applies a smaller current than CCCV. However, the increase in the SOC tracking error gradually requires a higher current to speed up. Meanwhile, internal resistance is reduced significantly along with the increase in SOC. Therefore, in the middle period, the charging current for MPC is larger than for CCCV. In the final stage, the current in the CV period is decreased to prevent over-voltage of the terminal voltage. Since the trajectory of the MPC is modified from the CCCV, the same trend is retained in MPC, keeping the terminal voltage under 4 V.

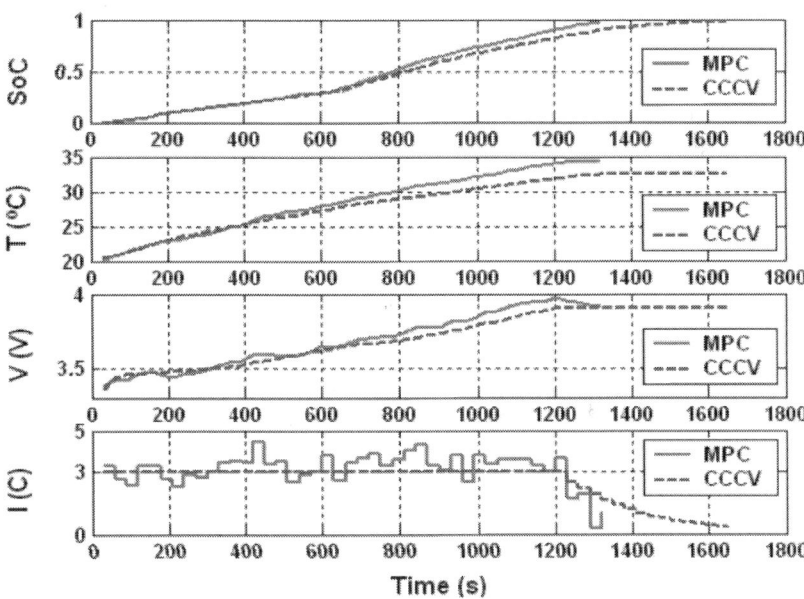

Figure 9. Curves during CCCV and MPC charging processes using modified 3 C profile.

5.4. Experimental Results

Since the maximum charging current of the programmable electric charger IT6154 is 9 A, in the experiment we use a single Li-ion cell with 2.3 Ah to demonstrate the fast charge performance of MPC. To protect the cells, we limit the charging current to below 3 C, *i.e.*, 6.9 A. The ambient temperature is   25 °C. The expected SOC trajectories of MPC are 1.10 times those of the CCCV results.

Figure 10 shows the Pareto-fronts of the CCCV and MPC charging results in experiments. All the results for MPC are superior to those for CCCV. Figure 11 illustrates the details of the 2 C charging processes. Unlike the simulation model, where the temperature is high in the lower SOC range of the specified battery type, the real LiFePO$_4$ cell has good charging acceptance at low SOC. Therefore, a high charging current is utilized to increase the charging speed. Before reaching around 80% SOC, another high charging period occurs. This is because MPC predicts the high temperature increase beyond 80%, so it applies a high charging current at this stage and switches to a low charging current when SOC is close to 1.

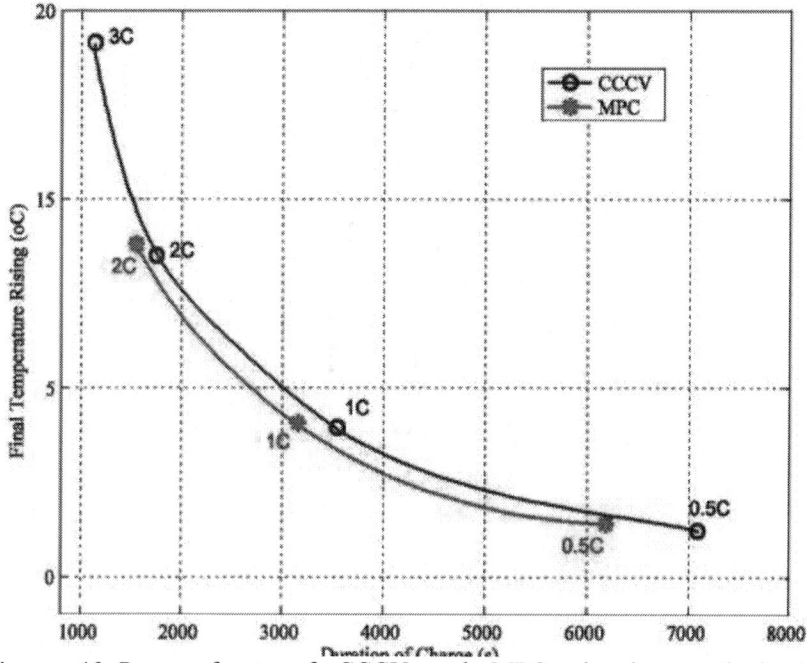

Figure 10. Pareto fronts of CCCV and MPC charging methods in experiments. The expected trajectories of MPC are modified from CCCV by multiplying by 1.10.

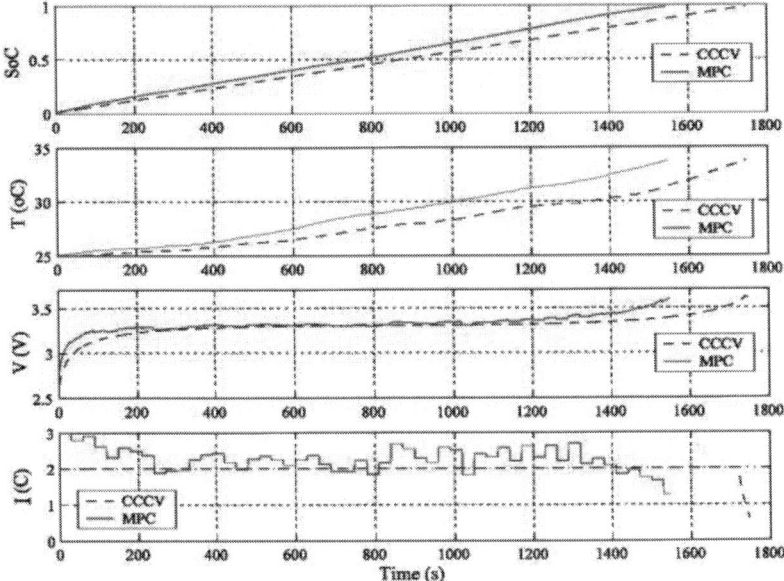

Figure 11. Experimental curves during the CCCV and MPC charging processes using a modified 2 C profile.

6. CONCLUSIONS

One of the most important obstacles for the mass commercialization of electric vehicles is the comparatively too slow process of refueling their energy. However, simply speeding up the charging process will increase the temperature rise, resulting in a low energy efficiency of charging process and maybe even damage to the batteries being charged. Therefore, the battery charge problem actually is a multi-objective issue which should be evaluated under the Pareto-optimal concept.

On the other hand, the charging stations of electric vehicles are usually designed as advanced equipment rather than a single chip based charger for portable devices. Thus, computers can be applied in charging stations to control the charging process. Although the proposed framework contains advanced algorithms such as NN and GA, it is practicable to be implemented on these computers, as shown in our experiment.

To accelerate charging speed and reduce the temperature increase, we introduce the MPC framework to the charging control process. Given a future charging sequence, the RC model and proposed modified ESC model are applied to predict the SOC in simulation and experiment, respectively. Meanwhile, besides the single-node lumped-parameter thermal model used in simulation, we also establish a neural network to model the thermal behavior of the applied battery in the experiment. A standard genetic algorithm is applied to optimize the charging current under the multi-objectives and constraints. The simulation and experimental results strongly demonstrate the availability and efficacy of MPC, with the conclusion that the Pareto front of the MPC dominates that of CCCV.

ACKNOWLEDGMENTS

This work was supported by the Shenzhen Key Laboratory of Electric Vehicle Powertrain Platform and Safety Technology.

REFERENCES

1. Cope, R.C.; Podrazhansky, Y. The Art of Battery Charging. In Proceedings of the 14th Annual Battery Conference on Applications and Advances, Long Beach, CA, USA, 12–15 January 1999; pp. 233–235.
2. Zhang, S.S.; Xu, K.; Jow, T.R. Study of the charging process of a $LiCoO_2$-based Li-ion Battery. *J. Power Sources* **2006**, *160*, 1349–1354.
3. Hsieh, G.C.; Chen, L.R.; Huang, K.S. Fuzzy-controlled Li-ion battery charge system with active state-of-charge controller. *IEEE Trans. Ind. Electron.* **2001**, *48*, 585–593.
4. Wong, Y.S.; Hurley, W.G.; Wolfle, W.H. Charge regimes for valve-regulated lead-acid batteries: Performance overview inclusive of temperature compensation. *J. Power Sources* **2008**, *183*, 783–791.
5. Notten, P.H.L.; Veld, J.H.G.; Beek, J.R.G. Boostcharging Li-ion batteries: A challenging new charging concept. *J. Power Sources* **2005**, *145*, 89–94.
6. Lin, C.H.; Hsieh, C.Y.; Chen, K.H. A Li-ion battery charger with smooth control circuit and built-in resistance compensator for achieving stable and fast charging. *IEEE Trans. Circuits Syst. I.* **2010**, *57*, 506–517.
7. Hussein, A.A.H.; Batarseh, I. A review of charging algorithms for nickel and lithium battery chargers. *IEEE Trans. Veh. Technol.* **2011**, *60*, 830–838.
8. Park, S.Y.; Miwa, H.; Clark, B.T.; Ditzler, D.S.K.; Malone, G.; D'souza, N.S.; Lai, J.S. A Universal Battery Charging Algorithm for Ni-Cd, Ni-MH, SLA, and Li-Ion for Wide Range Voltage in Portable Applications. In Proceedings of IEEE Conference on Power Electronics Specialists, Rhodes, Greece, 15–19 June 2008; pp. 4689–4694.
9. IKeya, T.; Sawada, N.; Murakami, J.; Kobayashi, K.; Hattori, M.; Murotani, N.; Ujiie, S.; Kajiyama, K.; Nasu, H.; Narisoko, H.; *et al.* Multi-step constant-current charging method for an electric vehicle

nickel/metal hydride battery with high-energy efficiency and long cycle life. *J. Power Sources* **2002**, *105*, 6–12.

10. Diaz, J.; Martin-Ramos, J.A.; Pernia, A.M.; Nuno, F.; Linera, F.F. Intelligent and universal fast charger for Ni-Cd and Ni-MH batteries in portable applications. *IEEE Trans. Ind. Electron.* **2004**, *51*, 857–863.

11. Cheng, M.W.; Wang, S.M.; Lee, Y.S.; Shiao, S.H. Fuzzy Controlled Fast Charging System for Lithium-Ion Batteries. In Proceedings of International Conference on Power Electronics and Drive Systems, Taipei, Taiwan, 2–5 November 2009; pp. 1498–1503.

12. Wang, J. Charging Strategy Studies for PHEV Batteries Based on Power Loss Model. In Proceedings of SAE World Congress & Exhibition, Detroit, MI, USA, April 2010. SAE Paper No: 2010-01-1238.

13. Ullah, Z.; Burford, B.; Dillip, S. Fast intelligent battery charging: Neural-fuzzy approach.*IEEE Aerosp. Electron. Syst. Mag.* **1996**, *11*, 26–34.

14. Khosla, A.; Kumar, S.; Aggarwal, K.K. Fuzzy Controller for Rapid Nickel-Cadmium Batteries Charger through Adaptive Neuro-Fuzzy Inference System (ANFIS) Architecture. In Proceedings of International Conference of the North American Fuzzy Information Processing Society, Chicago, USA, 24–26 July 2003; pp. 540–544.

15. Aliev, R.A.; Aliev, R.R.; Guirimov, B.; Uyar, K. Dynamic data mining technique for rules extraction in a process of battery charging. *Appl. Soft Comput.* **2008**, *8*, 1252–1258.

16. Liu, Y.H.; Teng, J.H.; Lin, Y.C. Search for an optimal rapid charging pattern for lithium-ion batteries using ant colony system algorithm. *IEEE Trans. Ind. Electron.* **2005**, *52*, 1328–1336.

17. Saberi, H.; Salmasi, F. Genetic Optimization of Charging Current for Lead-Acid Batteries in Hybrid Electric Vehicles. In Proceedings of International Conference on Electrical Machines and Systems, Seoul, Korea, 8–11 October 2007; pp. 2028–2032.

18. Chen, L.R.; Hsu, R.C.; Liu, C.S. A design of a grey-predicted Li-ion battery charge system. *IEEE Trans. Ind. Electron.* **2008**, *55*, 3692–3701.

19. Salamch, Z.M.; Casacca, M.A.; Lynch, W.A. A mathematical model for lead-acid batteries. *IEEE Trans. Energy Convers.* **1992**, *7*, 93–98.

20. Bergveld, H.J.; Kruijt, W.S.; Notten, P.H.L. *Battery Management System: Design by Modeling*; Kluwer Academic Publishers: Dordrecht, The Netherlands, 2002.

21. Johnson, V.H. Battery performance models in ADVISOR. *J. Power Sources* **2002**, *110*, 321–329.

22. Plett, G.L. Extended Kalman filtering for battery management systems of LiPB-based HEV battery packs: Part 2. Modeling and identification. *J. Power Sources* **2004**, *134*, 262–276.

23. Szumanowski, A.; Chang, Y. Battery Management System Based on Battery Nonlinear Dynamics Modeling. *IEEE Trans. Veh. Technol.* **2008**, *57*, 1425–1432.
24. Camacho, E.F.; Bordons, C. *Model Predictive Control*; Springer Verlag: London, UK, 2004.
25. Qin, S.J.; Badgwell, T.A. A survey of industrial model predictive control technology.*Control Eng. Pract.* **2003**, *11*, 733–764.
26. Bemporad, A.; Morari, M. Robust Model Predictive Control: A Survey. In *Robustness in Identification and Control*; Springer Verlag: London, UK, 1999; pp. 207–226.
27. Yan, J.Y.; Ling, Q.; Chen, W. Nonlinear Model Predictive Control Based on Evolutionary Algorithms: Framework, Theory and Application. In *A Giordano and G. Costa, Soft Computing: New Research*; Alessia, J.G., Ginevra, E.C., Eds.; Nova Publishers: New York, NY, USA, 2009.
28. Pesaran, A.A. Battery thermal models for hybrid vehicle simulations. *J. Power Sources***2002**, *110*, 377–382.
29. Chow, M.; Lukic, S.; Wang, L.; Govindaraj, A. Research report on A123 battery modeling. Available online: http://www.adac.ncsu.edu/projects/Battery%20Model/Docs/Resea rch%20Repot%20on%20A123%20Battery%20Modeling.ppt (accessed on 4 August 2011).
30. Plett, G.L. Extended Kalman filtering for battery management systems of LiPB-based HEV battery packs-Part 3. State and parameter estimation. *J. Power Sources* **2004**, *134*, 277–292.
31. Plett, G.L. Sigma-point Kalman filtering for battery management systems of LiPB-based HEV Battery packs: Part 1: Introduction and state estimation. *J. Power Sources* **2006**, *161*, 1356–1368.
32. Plett, G.L. Sigma-point Kalman filtering for battery management systems of LiPB-based HEV Battery packs: Part 2: Simultaneous state and parameter estimation. *J. Power Sources***2006**, *161*, 1369–1384.
33. Yan, J.Y.; Xu, G.Q.; Xie, B. Battery State-of-Charge Estimation Based on H Filter for Hybrid Electric Vehicle. In Proceedings of International Conference on Control, Automation, Robotics and Vision, Hanoi, Vietnam, 17–20 December 2008; pp. 464–469.
34. Deb, K. *Multi-Objective Optimization Using Evolutionary Algorithms*; Wiley: Hoboken, NJ, USA, 2001.
35. Jiang, B.; Wang, B.W. Parameter estimation of nonlinear system based on genetic algorithm. *Control Theory Appl.* **2000**, *17*, 150–152.
36. Goldberg, D.E.; Deb, K. A comparative analysis of selection schemes used in genetic algorithms. *Found. Genet. Algorithms* **1991**, *1*, 69–93.

CHAPTER 10
State Space Model Predictive Control of an Aerothermic Process with Actuators Constraints

Mustapha Ramzi[1], Hussein Youlal[2], Mohamed Haloua[3]

[1]ASTIMI Laboratory, High School of Technology, Université Mohammed V Agdal, Rabat, Morocco
[2]UFR Automatic and Information Technologies, Faculty of Science, Rabat, Morocco
[3]Automatic Laboratory, Mohammadia School of Engineering, Rabat, Morocco

ABSTRACT

This paper investigates State Space Model Predictive Control (SSMPC) of an aerothermic process. It is a pilot scale heating and ventilation system equipped with a heater grid and a centrifugal blower, fully connected through a data acquisition system for real time control. The interaction between the process variables is shown to be challenging for single variable controllers, therefore multi-variable control is worth considering. A multi-variable state space model is obtained from on-line experimental data. The controller design is translated into a Quadratic Programming (QP) problem, in which a cost function subject to actuators linear inequality constraints is minimized. The outcome of the experimental results is that the main control objectives, such as set-point tracking and perturbations rejection under actuators constraints, are well achieved for both controlled variables simultaneously.

Keywords: Multi-Variable Control; Aerothermic Process; Actuators Constraints; Process Identification; State Space Model Predictive Control

1. INTRODUCTION

The heating and ventilation system plays an important role in our daily life where certain desired temperature is controlled in order to maintain the healthy and safe working environment to the conditioned space. It is also the case in many industrial sectors including chemical, mineral, drying and distillation processes, as well as pharmaceutical and agroalimentary production units. It is argued that the temperature control is no more a challenging control problem in most of these applications. Nevertheless, some practical issues in many temperature control applications stimulate new developments and farther investigations [1-4].

For education and training purposes many aerothermic processes are available. They highlight most heating and ventilation problems, and they are widely referenced in process control literature. Different prototypes of these processes have been used to check new control strategies and many results were reported for the single variable control cases [5-9]. The aerothermic processes have generally a thermal protection for which they are entirely stopped when electrical power is maximal and the ventilator speed signal is under a given threshold.

In addition to these physical limits, there exists a significant interaction between the main processes variables which results from the nonlinearity of the process as reported in [8]. However, these constraints were not explicitly considered in most reported control approaches for aerothermic processes. Hence, the design of a multivariable feedback control system is worth to investigate. Among the many valuable approaches to face this kind of control problems, the Model Predictive Control (MPC) with constraints has been considered in this work. This choice is motivated by the fact that the MPC control has been investigated and successfully employed in some complex industrial processes [10-16].

In this paper, the State Space Model Predictive Control (SSMPC) with actuators constraints is considered for a pilot scale aerothermic process. To fulfil the requirement for integral action in most industrial control systems, we have embedded the SSMPC design model with integrators to achieve this objective and ensure outputs steadystate error free. This strategy is transformed into a Quadratic Programming (QP) problem, in which a quadratic cost function subject to linear inequality constraints is minimized on-line. The implementation of the predictive control in real time is based on the result of this minimization and only the first input of the optimal command sequence is used each time a new state is updated. In the synthesis of the SSMPC controller, a state space model is identified using the Numeric Subspace State Space System IDentification (N4SID). This technique has attracted an increasing attention of several researchers in the last few years [17-22]. It provides a robust and accurate method for the identification of dynamical systems under the influence of perturbations. Among the advantages of the N4SID method, we mention its ability to deal with multi-input multi-output identification in a straightforward manner from process experimental data and the ease of use due to the small number of parameters which have to be chosen by the user. This is a method that does not require nonlinear searches in the parameter space, but it is based only on computational tools such as the QR factorisation, and the singular-value decomposition (SVD), which make it robust and numerically stable [20]. The method contrasts with the robust design used in [1]. In this paper, we examine various issues of both N4SID identification and SSMPC control performances achieved experimentally on a pilot scale aerothermic process. The objectives of the proposed control technique which are about reaching reference set-points for the temperature and the air flow, subject to effects of both actuators constraints and the external perturbations. These goals are achieved by manoeuvring the heating resistance and the ventilator speed under constraints on the manipulated variables and their rate of change to handle the factory set thermal protection. Worth to mention herein that the basic factory control system delivered with the process is restricted to classical analog PID control, and most reported literature work on this kind of process deal with mainly mono-variable digital control. The results reported herein highlight further aspects of multivariable control of the considered process.

The paper content is organised as follows: Section 2 introduces the description of the aerothermic process and underlines the interaction between the main process variables. Section 3 discusses the multivariable state space identification, which is the first step in the design of the controller. Section 4 introduces the SSMPC algorithm where integral actions and set-point tracking are naturally embedded in the algorithm. In this section, we recall the main steps in the development of quadratic programming which implement the SSMPC. Section 5 reports the experimental control results of the aerothermic process operation under various inputs perturbations. Robustness of the SSMPC controller is also discussed and a final conclusion is given.

2. AEROTHERMIC PROCESS DESCRIPTION

The considered pilot scale aerothermic process [23], is shown as a schematic diagram in **Figure 1**and depicted in a three dimensional view in **Figure 2**. As described in [24], it has the basic characteristics of a large process, with a tube through which atmospheric air is drawn by a centrifugal blower, and is heated as it passes over a heater grid before being released into the atmosphere.

The command objective for the aerothermic process is to regulate the temperature and the air flow by guaranteeing the verification of the full actuators constraints. The temperature control is achieved by varying the electrical power supplied to the heater grid. There is an energized electric resistance inside the tube, and due to the Joule effect, heat is released by the resistance and transmitted, by convection, to the circulating air, resulting in heated air. The air flow is adjusted by varying the speed of the fan.

This process can be characterized as a non-linear system. The physical principle which governs the behaviour of the aerothermic process is the balance of heat energy. Hence, when the air temperature and the flow inside the process are assumed to be uniform, a linear system model can be obtained.

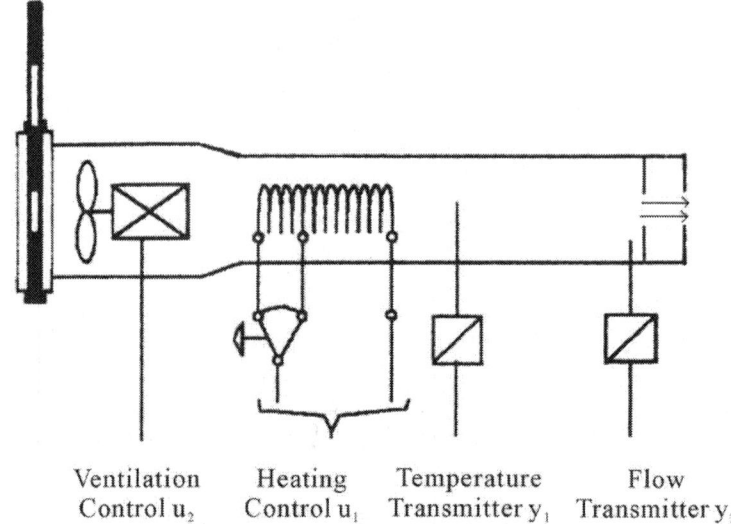

Figure 1. Schematic illustration of aerothermic process.

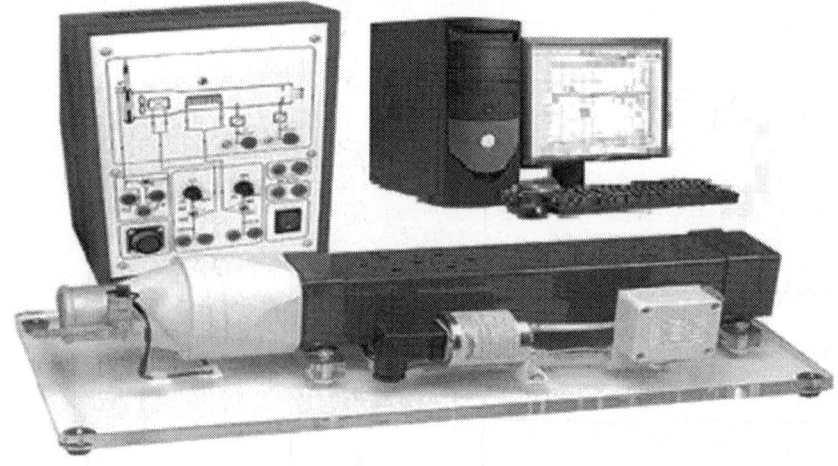

Figure 2. Three-dimensional view of aerothermic process.

As shown in the schematic of the aerothermic process, the system inputs, (u_1, u_2), are respectively the power electronic circuit feeding the heating resistance and the ventilator speed. The outputs, (y_1, y_2), are respectively the flow and air temperature. The input-output signals are expressed by a voltage, between 0 and 10 V, issued from the transducers and conditioning electronics.

To examine the possibility of interaction between the temperature and air flow of the aerothermic process, two experiments were carried out. In each case, the two process inputs were held constant and allowed to settle. If one of them undergoes a step change, the behaviour of the other output will be observed to see if this change had any effect on it. **Figure 3** shows the results from both experiments. In the first half plot, the electric voltage supplied to the heater grid is held constant (at 4 V) and the speed of the fan undergoes a step change from 30% to 70% of its full range. The air temperature varied considerably from 4 V (45°C) to 2 V (35°C). The second half plot shows the results when the fan speed is held constant and the electric voltage of the heater grid undergoes a step change, from 40% to 80% of full range. As can be seen, the air temperature is varied accordingly but the air flow is remained unaffected. These results show that the air temperature behaviour depends also strongly on the operating conditions of the air flow.

3. STATE SPACE IDENTIFICATION

System identification is an experimental approach to determine the transfer function or equivalent mathematical description for the dynamic of an industrial process component by using a suitable input signal. This approach represents the first step in the design of a controller.

A considerable number of system identification methods have been investigated and they are generally classified into parametric approaches. In contrast to these classical algorithms, the State Space Method Identification (N4SID) does not suffer from the problems caused by a priori parameterizations and non-linear optimisations. They identify MIMO systems in a very simple and elegant way. Among his advantages we mention: these ability to deal with multi-input multi-output in a straightforward manner from process experimental data and the ease of use due to the small number of parameters which have to be chosen by the user. They are methods which do not require nonlinear searches in the parameter space, but it is based only on computational tools such as the QR and the singular-value decomposition (SVD), which make it robust and numerically stable [20].

In order to generate estimation and validation data for system identification, an experiment is performed. Data set used for the parameter identification step is build up with Pseudo Random Binary Sequence (PRBS) signals which are applied simultaneously to the two manipulated variables. This data set is displayed in **Figure 4**.

The sampling interval is $T_s = 1$ second. The signals collected, via the MF624 data acquisition module, are yield in the interval (0 V, 10 V). After the application of N4SID algorithm on first half experimental data of identification (i.e.: 100 minutes), the model of the aerothermic process is given by the following discrete state-space representation:

$$\begin{cases} x_p(k+1) &= A_p x(k+1) + B_p x(k) \\ y(k) &= C_p x(k) + D_p u(k) \end{cases}$$

$$(1)$$

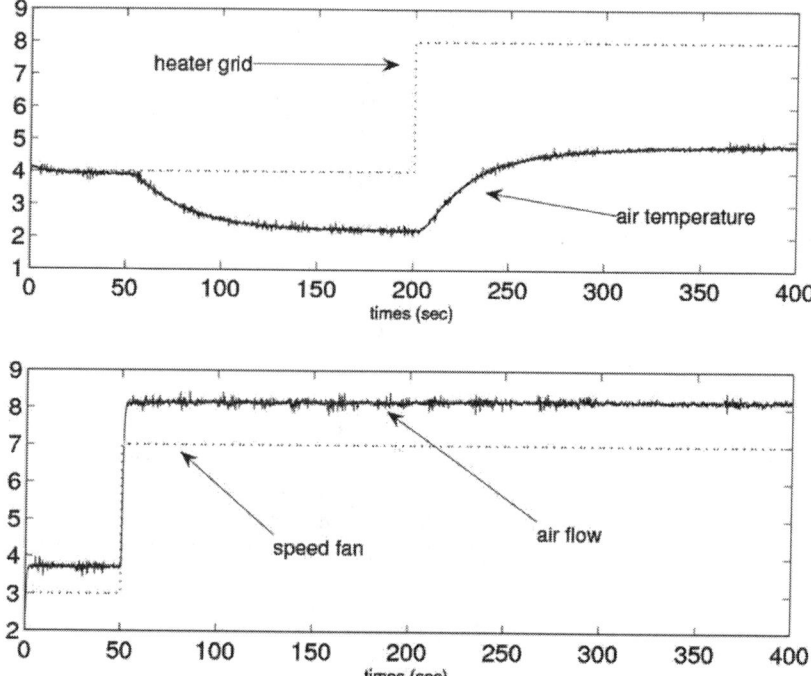

Figure 3. Interaction between the aerothermic process main variables.

with

$$A_p = \begin{pmatrix} 0.9819 & -0.0024 & 0.0009 & -0.189 \\ 0.0800 & 0.5159 & 0.2760 & 0.0679 \\ 0.0270 & -0.6286 & -0.2750 & -0.2292 \\ 0.0810 & -0.0442 & -0.3830 & 0.7457 \end{pmatrix}$$

$$B_p = \begin{pmatrix} 0.0005 & -0.0003 \\ -0.0001 & 0.0256 \\ 0.0018 & 0.1002 \\ -0.0085 & 0.0222 \end{pmatrix}$$

$$C_p = \begin{pmatrix} 25.2618 & 1.0062 & -0.8844 & 2.4627 \\ -2.2614 & 14.8776 & -3.5444 & -1.2790 \end{pmatrix}$$

The matrix D_p is equal to zero, $u = [u_1, u_2]^T$ and $y = [y_1, y_2]^T$.

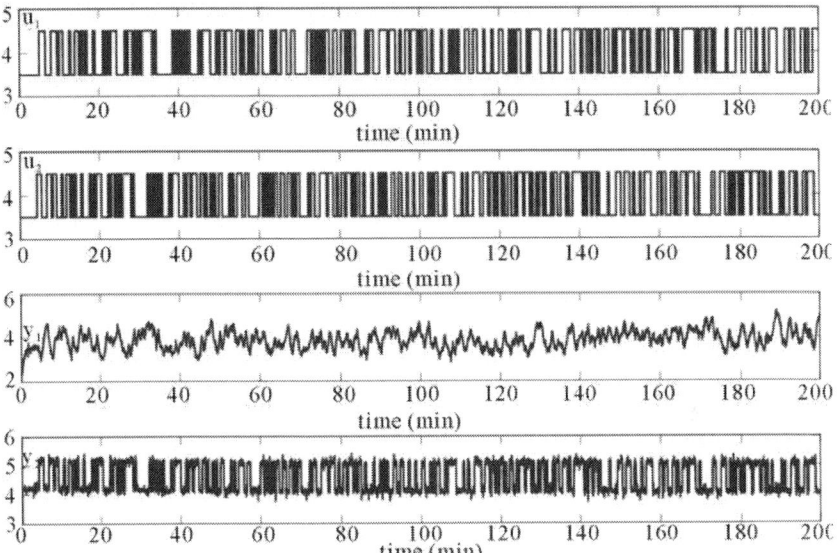
Figure 4. Data set for state space identification.

The system described by these matrices is stable, completely observable and controllable.

Once the model is identified, we have validated it by comparing his estimate output, when the input of the remaining experimental data is applied, with the true system output. This comparison is represented by the **Figure 5**.

As shown in this figure, it appears a good similarity between the true system output and the identified one.

4. CONTROL PROBLEM AND SSMPC FORMULATION

The deterministic model of aerothermic process be controlled has two inputs, two outputs and four states. When the plant noise and perturbation are taken into consideration, the Equation (1) describing the aerothermic process becomes:

$$\begin{cases} x_p(k+1) = A_p x(k+1) + B_p x(k) + B_d w(k) \\ y(k) = C_p x(k) + D_p u(k) + D_d w(k) \end{cases}$$

(2)

where $x_p(k)$ is the (4×1) state plant vector, $u(k)$ is the (2×1) control input vector, $y(k)$ is the (2×1) process output vector and $w(k)$ is a (2×1) vector of perturbations. The matrices D_u and D_d are assumed to be zero, this imply that there is not direct feed through of the manipulated variable and the perturbations on the output vector. A_p, B_p, B_d and C_p are matrices of appropriate dimensions. In order to ensure that integrators are embedded in the identified model, we need to change the model to suit this design purpose as in [10]. Taking a difference operation on both sides of the state equation in (2) yields:

$$x_p(k+1) - x_p(k) = A_p(x_p(k) - x_p(k-1))$$
$$+ B_p(u(k) - u(k-1)) + B_d(w(k) - w(k-1)) \tag{3}$$

In general, the incremental of the variable v(k) is de-noted by

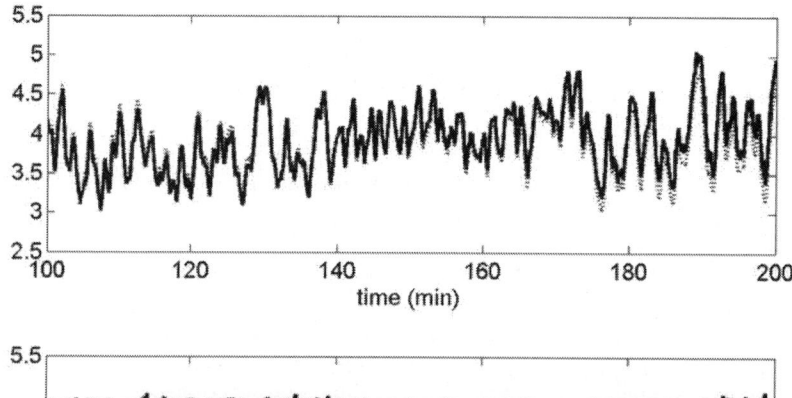

Figure 5. Outputs (solid line) and their estimates (dotted line).

$$\Delta v(k) = v(k) - v(k-1)$$

With this information, the incremental state-space equation can be written as:

$$\Delta x_p(k+1) = A_p \Delta x_p(k) + B_p \Delta u(k) + B_d \varepsilon(k) \tag{4}$$

where $\Delta u(k)$ is the input to the state-space model, also called the rate of change of the control inputs, and $\varepsilon(k) = w(k) - w(k-1)$.

In order to relate the output y(k) to the state variable $\Delta x_p(k)$, a new state variable vector is chosen to be

$$x(k) = \left[\Delta x_p(k)^T, y(k) \right]^T$$

where superscript T indicates matrix transpose.
From (4) we deduce that

$$y(k+1) = y(k) + C_p A_p \Delta x_p(k)$$
$$+ C_p B_p \Delta u(k) + C_p B_p \varepsilon(k) \tag{5}$$

Putting together (4) and (5) leads to the following augmented state-space model:

$$\left\{ \begin{array}{l} \begin{bmatrix} \Delta x_p(k+1) \\ y(k+1) \end{bmatrix} = \begin{bmatrix} A_p & 0_p^T \\ C_p A_p & I_{q\times q} \end{bmatrix} \begin{bmatrix} \Delta x_p(k) \\ y(k) \end{bmatrix} \\[1em] \quad + \begin{bmatrix} B_p \\ C_p B_p \end{bmatrix} \Delta u(k) + \begin{bmatrix} B_d \\ C_p B_d \end{bmatrix} \varepsilon(k) \\[1em] \quad y(k) = \begin{bmatrix} 0_{q\times n} & I_{q\times q} \end{bmatrix} \begin{bmatrix} \Delta x_p(k) \\ y(k) \end{bmatrix} \end{array} \right. \tag{6}$$

where the subscripts q and n are respectively the number of outputs and the state space model dimension. $0_{q\times n}$ is a q × n zero matrix and $I_{q\times q}$ the q × q identity matrix.

With the assumption that $\varepsilon(k)$ is a zero-mean white noise sequence, the predicted value i samples ahead $\varepsilon(k + i)$ is assumed to be zero.

For notational simplicity, we rewrite the augmented state-space model (6) as:

$$\left\{ \begin{array}{l} x(k+1) = Ax(k) + B\Delta u(k) \\ \quad\quad y(k) = Cx(k) \end{array} \right. \tag{7}$$

where A, B and C are matrices corresponding to (6). Their computation for the identified state space model of the considered aerothermic process yields the following results:

$$A = \begin{pmatrix} 0.9819 & -0.0024 & 0.0009 & -0.0189 & 0 & 0 \\ 0.0800 & 0.5159 & 0.2760 & 0.0679 & 0 & 0 \\ 0.0270 & -0.6286 & -0.2750 & -0.2292 & 0 & 0 \\ 0.0810 & -0.0442 & -0.3830 & 0.7457 & 0 & 0 \\ 25.0614 & 0.9046 & -0.3986 & 1.6305 & 1 & 0 \\ -1.2297 & 0.9652 & 5.5689 & 0.9110 & 0 & 1 \end{pmatrix}$$

$$B = \begin{pmatrix} 0.0005 & -0.0003 \\ -0.0001 & 0.0256 \\ 0.0018 & 0.1002 \\ -0.0085 & 0.0222 \\ -0.0111 & -0.0168 \\ 0.0024 & -0.0019 \end{pmatrix}, \quad C = \begin{pmatrix} 0 & 0 & 0 & 0 & 1 & 0 \\ 0 & 0 & 0 & 0 & 0 & 1 \end{pmatrix}$$

The eigenvalues of the augmented model are given by:

$$\lambda = \begin{pmatrix} 1 \\ 1 \\ -0.0208 \\ 0.1774 \\ 0.8423 \\ 0.9694 \end{pmatrix}$$

The first two components of λ are from the augmentation of the state space model, and the last four are from the original plant. This means that there are 2 integrators embedded into the augmented design model, which ensures integral action for the SSMPC controller.

Define the vector Y and ΔU as

$$\begin{cases} \Delta U = \left[\Delta u(k)^T \ \Delta u(k+1)^T \cdots \Delta u(k+N_c-1)^T \right] \\ Y = \left[y(k+1)^T \ y(k+2)^T \cdots y(k+N_p)^T \right] \end{cases}$$

$$(8)$$

where N_p denotes the length of the prediction horizon or output horizon, and N_c denotes the length of the control horizon or input horizon.

Based on the state-space model (7), the future state variables are calculated sequentially using the set of future control parameters. After calculating the predicted output variables, we have the following compact matrix

$$Y = Fx(k) + \Phi \Delta U$$

(9)

where

$$F = \begin{bmatrix} C \\ CA \\ ... \\ CA^{N_p} \end{bmatrix}$$

$$\Phi = \begin{bmatrix} CB & 0 & 0 & ... & 0 \\ CAB & CB & 0 & ... & 0 \\ CA^2 B & CAB & CB & ... & 0 \\ . & . & . & . & . \\ CA^{N_p-1}B & CA^{N_p-2}B & CA^{N_p-3}B & ... & CA^{N_p-N_c}B \end{bmatrix}$$

At time k, the state variable x(k) is not measurable (i.e.: C is different to the identity matrix). The control law is computed using the estimated state variables given by the following equation:

$$\hat{x}(k+1) = A\hat{x}(k) + B\Delta u(k) + K_{obs}\left(y(k) - C\hat{x}(k)\right)$$

where K_{obs} is the Kalman filter gain Obtained by solving recursively (for i = 0, 1,···) the following equation:

$$K_{obs}(i) = AP(i)C^T\left(\alpha + CP(i)C^T\right)^{-1}$$
$$P(i+1) = Ap(i)A^T - AP(i)C^T$$
$$\left(\alpha + CP(i)C^T\right)^{-1}CP(i)A^T + \beta$$

where α and β are the matrices to be chosen by the user.

For a given set-point signal r(k) at sample time k, the main control objective is to bring the predicted output as close as possible to the set-point signal and annulled the effect of the perturbations with respect the actuators constraints. This objective is translated into a design to find the control parameter vector ΔU such that an error function between the set-point signal and the predictive output is minimized. The cost function J that reflects the control objective is defining as:

$$J = \Delta U^T \left(\phi^T \phi + R \right) \Delta U - 2 \Delta U^T \phi^T \left(R_s - Fx(k) \right) \tag{10}$$

Subject to the inequality constraints

$$M \Delta U \leq \gamma$$

where M is a matrix reflecting the constraints and

$$R_s^T = I_{N \, pxq} \, r(k)$$

The matrix $\Phi^T\Phi$ has dimension $mN_c \times mN_c$ and Φ^TF has dimension $mN_c \times n$ and Φ^TR equals the last q columns of Φ^TF. The weight matrix R is a block matrix with m blocks and has its dimension equal to the dimension of $\Phi^T\Phi$.

Since the cost function (10) is a quadratic, and the constraints are linear inequalities, the problem of funding an optimal predictive control becomes on of finding an optimal solution to standard Hildreth's quadratic programming problem [10,11] and [25]. Hence, the problem is written as minimizing

$$J = 1/2 \, \chi^T \varsigma \chi + \chi^T \psi \tag{11}$$

Subject to the inequality constraints

$$M \chi \leq \gamma \tag{12}$$

Where

$\varsigma = 2 \left(\phi^T \phi + R \right)$ and $\psi = -2\phi^T \left(R_s - Fx(k) \right)$

One the optimal solution to (11) at time k is obtained on line, its first element is applied to (1). The optimization (11) is repeated at time k + 1, based on the new state $\hat{x}_p(k+1)$, yielding a moving horizon control strategy.

5. EXPERIMENTAL RESULTS

The objectives of the control technique applied to aerothermic process are summarized below:

- The temperature and the air flow must reach given reference set-points.
- The actuators constraints must be verified and respected.
- The effect of the perturbations must be annulled.
- The effect of the interaction caused by the speed ventilator on the air temperature must be eliminated.

The predictive control setup described in the previous section was first tested in simulation using the model obtained from identification. This investigation was done especially to evaluate the computational complexity of the controller and to find the N_c and N_p before the application of the controller in the real aerothermic process. Hence, we find $N_c = 3$ and $N_p = 20$ and the weight matrix R $= 0.5I_{6\times6}$.

The implementation of the SSMPC, in real time, uses the Humusoft MF624 Data Acquisition Card of 14-bit Analog to Digital (A/D) conversion module, plugged into ISA port. The signals are transmitted between the PC and the Aerothermic Process via a 37-way cable and connector block.

In this experimental study, two types of echelon perturbations can be envisaged in order to challenge the control performances. The first one is characterized by the rotating of the diaphragm to 90 degrees. This perturbation affects the temperature and the air flow outputs. The second perturbation is characterized by the changing of the switch position (S) of the resistance heating. This perturbation affects the heater grid value, which increases the air temperature. The **Figure 6** represents the aerothermic process controlled by the MIMO SSMPC technique. y, u, r and ω represent respectively the measured output or controlled variable, the manipulated input, the set-point and the perturbations. These vectors can be written as follows:

$$ y = \begin{bmatrix} y_1 \\ y_2 \end{bmatrix}, u = \begin{bmatrix} u_1 \\ u_2 \end{bmatrix}, r = \begin{bmatrix} r_1 \\ r_2 \end{bmatrix}, w = \begin{bmatrix} w_1 \\ w_2 \end{bmatrix} $$

where w_1 is the opening of the diaphragm and w_2 is the heater grid level.

The constraints on the manipulated variables u(k) and their rate of change Δu(k) are taken into account to accommodate the system thermal protection for which the aerothermic process is entirely stopped when electrical power is maximal and the ventilator speed signal is under a threshold of about 1.25 V. Based on practical considerations of the process operation, these constraints may be summarized as follows:

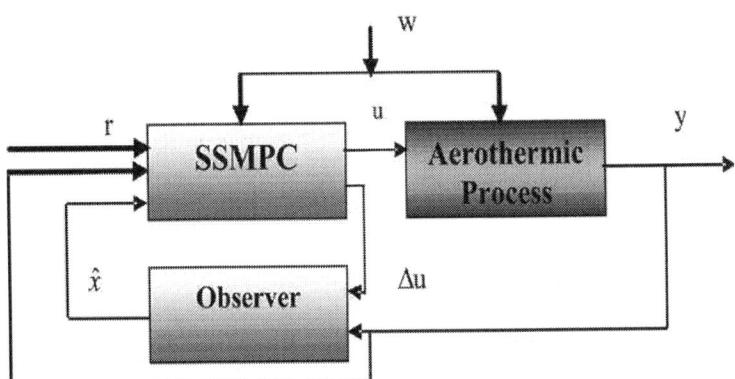

Figure 6. Block diagram of MIMO SSMPC application.

$$1.25 \text{ V} \le u(k) \le 7 \text{ V} \quad \text{and} \quad -1 \text{ V} \le \Delta U \le 1 \text{ V}$$

Furthermore the set of linear inequality constraints given by (12) is formulated into the following matrix form:

$$M = \begin{bmatrix} M_1 \\ M_2 \\ -M_2 \end{bmatrix}, \qquad \chi = \begin{bmatrix} \Delta u_1(k) \\ \Delta u_2(k) \\ \Delta u_1(k+1) \\ \Delta u_2(k+1) \\ \Delta u_1(k+2) \\ \Delta u_2(k+2) \end{bmatrix}$$

$$\gamma = \begin{bmatrix} I_{6\times1} \\ I_{6\times1} \\ (7 - u(k-1)) I_{6\times1} \\ (-1.25 + u(k-1)) I_{6\times1} \end{bmatrix}$$

where $I_{6\times1}$ is a column vector of ones and $I_{6\times6}$ denotes the identity matrix, $M_1 = [I_{6\times6}, -I_{6\times6}]^T$,

$$M_2 = \begin{bmatrix} 1 & 0 & 0 & 0 & 0 & 0 \\ 0 & 1 & 0 & 0 & 0 & 0 \\ 1 & 0 & 1 & 0 & 0 & 0 \\ 0 & 1 & 0 & 1 & 0 & 0 \\ 1 & 0 & 1 & 0 & 1 & 0 \\ 0 & 1 & 0 & 1 & 0 & 1 \end{bmatrix}.$$

In our experimental application, the computation of K_{obs} gives the following result:

$$K_{obs} = \begin{bmatrix} 0.0382 & -0.0025 \\ 0.0057 & 0.0082 \\ -0.0025 & -0.0118 \\ 0.0053 & 0.0094 \\ 1.9866 & -0.0217 \\ 0.0004 & 1.1572 \end{bmatrix}$$

As shown in **Figure 7**, the set-points of the temperature and the air flow are changed respectively at 10 and 20 minutes. From this figure, it can be observed that both the temperature and the air flow reach their set-points imposed by respecting the operational full actuator constraints. The Figures 8 and 9 show the associated control signal responses and rates of change on both control signals respectively. The **Figure 8** shows clearly the comportment of the control u_1 at time 20 minutes in order to maintain the temperature at his desired set-point when the air flow reference has been changed. What means that the aerothermic process variables are coupled.

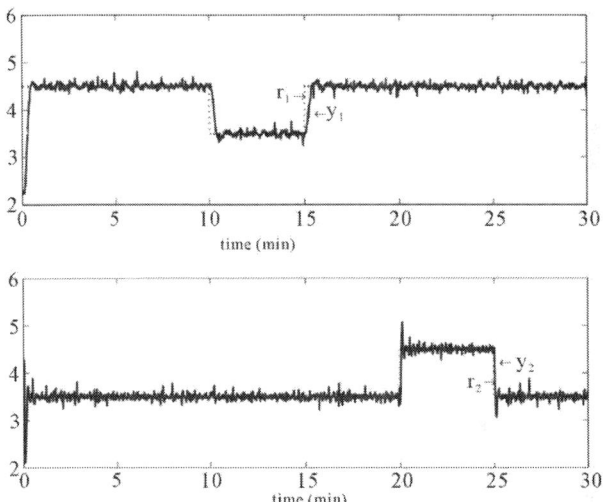

Figure 7. Closed-loop response: air temperature (top figure); air flow (bottom figure).

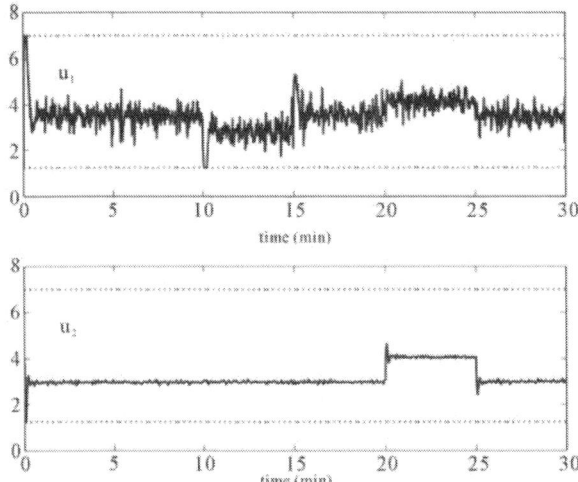

Figure 8. Closed-loop control signal response with actuator constraints. Top figure: heater grid ($1.25 \leq u1(k) \leq 7$); bottom figure: Ventilator speed ($1.25 \leq u2(k) \leq 7$).

The **Figure 10** shows the output responses when the perturbations are affected u_1 and u_2 respectively; while Figures 11 and 12 show the associated control signal responses and rates of change on both control signals respectively. As shown in the **Figure 10**, the perturbation on the air flow caused by opening of the diaphragm to 90 degrees, at time 31.75 minutes of the experience, is completely rejected. The perturbation on the air temperature, at time 42.8 minutes is also rejected. These rejections are due to the two integrators effect embedded in the SSMPC controller. The **Figure 11** shows clearly the behaviour of the two command variables towards this rejection.

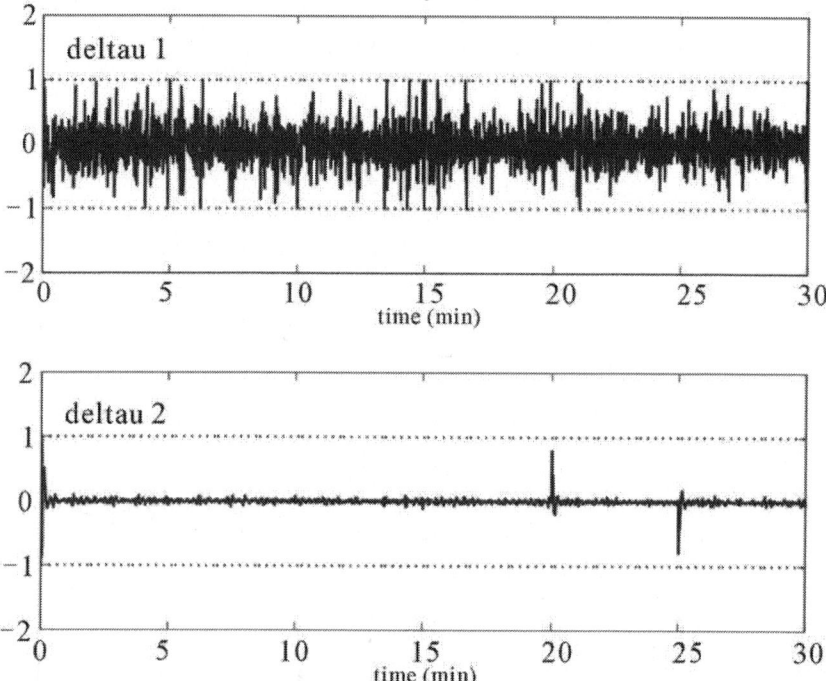

Figure 9. Rate of changes in control signal response. Top figure: rate of change for heater grid ($-1 \leq \Delta u1(k) \leq 1$); bottom figure: Rate of change for ventilator speed ($-1 \leq \Delta u2(k) \leq 1$).

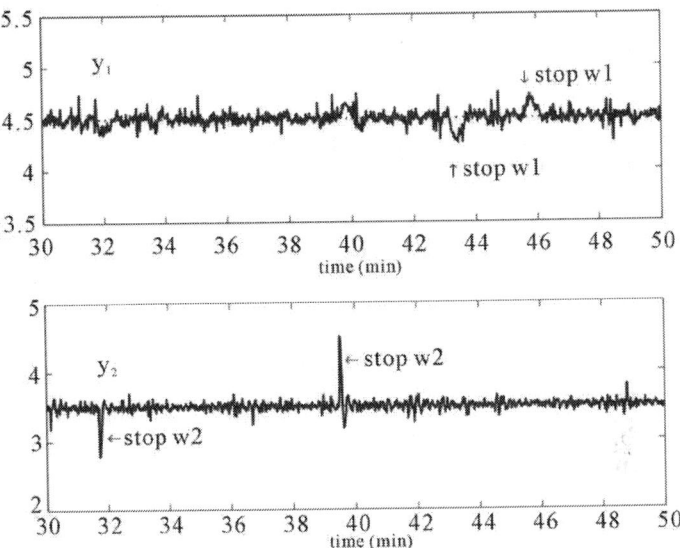

Figure 10. Closed loop response with perturbations. Top figure: air temperature; bottom figure: air flow.

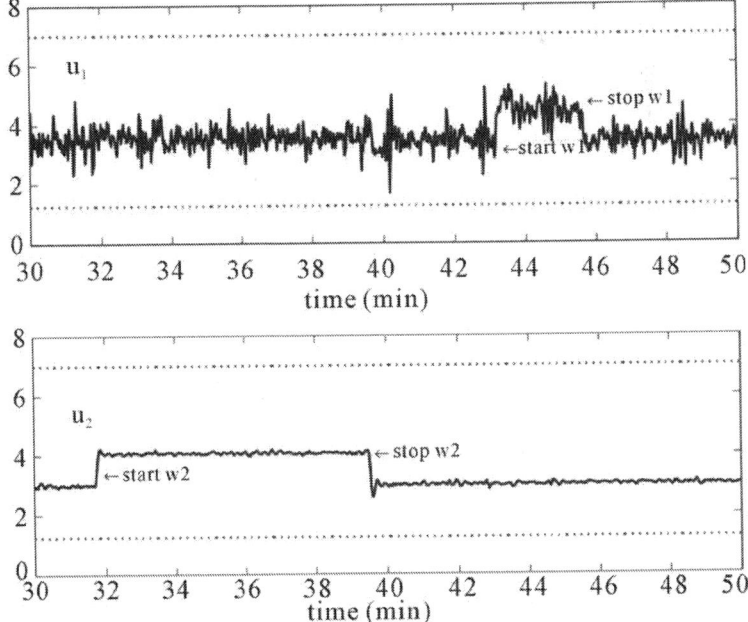

Figure 11. Closed loop control signal response with actuator constraints perturbations. Top figure: heater grid ($1.25 \leq u1(k) \leq 7$); bottom figure: Ventilator speed ($1.25 \leq u2(k) \leq 7$).

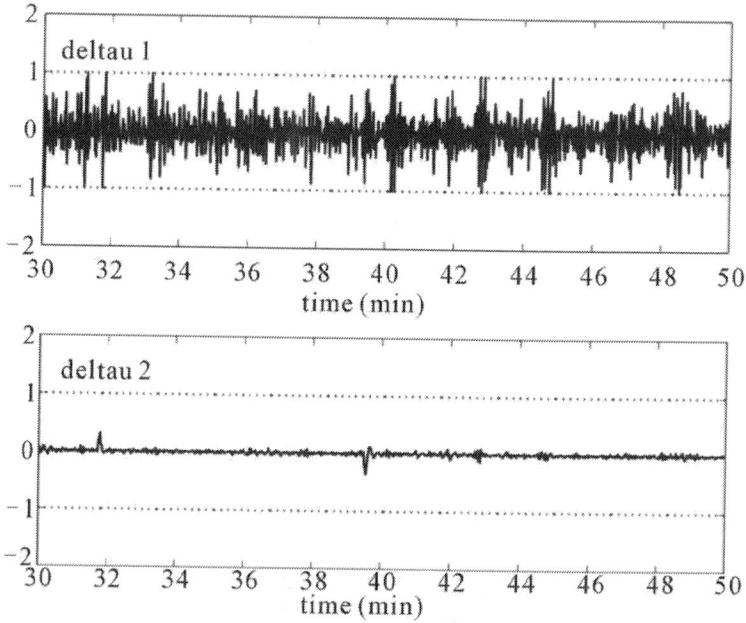

Figure 12. Rate of changes in control signal response. Top figure: rate of change for heater grid $(-1 \leq \Delta u1(k) \leq 1)$; bottom figure: rate of change for ventilator speed $(-1 \leq \Delta u2(k) \leq 1)$.

These experimental results show the efficiency of the SSMPC strategy, proposed in this paper, to control the temperature and air flow of the aerothermic process regardless the possible mismatch between the nonlinear process and his identified model. Furthermore, the SSMPC optimization problem has well taken into account the full actuators constraints compatible with the aerothermic process instruments. Noting that, the plant output constraints can be used. But, they are not required in this experimental application.

6. CONCLUSION

In this paper we have described a predictive control design approach with embedded integrators for a pilot scale aerothermic process. The number of these integrators is set equal to the number of outputs which make them steady-state error free. The proposed control approach is versatile in that it allows embedding full actuators constraints. The SSMPC application results completely satisfy the requested specifications. The State Space Method Identification (N4SID) is used to identify the basic model of the SSMPC controller. An observer based on the Kalman filter is used to estimate the aerothermic process state variable. In conclusion, the use of MIMO SSMPC demonstrates robust performance for tracking set point changes and rejecting the perturbations without violating constraints. It constitutes a worth extension of the monovariable control methods and an alternative to the basic classical control for these kind of processes used for both engineering education and research training.

ACKNOWLEDGEMENTS

The authors would like to thank Professor Liuping Wang (School of Electrical and Computer Engineering, Royal Melbourne Institute of Technology (RMIT) University, Australia) for her very helpful remarks and suggestions to achieve this document.

REFERENCES

1. N. Bennis, J. Duplaix, G. Enéa, M. Haloua and H. Youlal, "Greenhouse Climate Modelling and Robust Control," Computers and Electronics in Agriculture, Vol. 61, No. 2, 2008, pp. 96-107.

2. M. Nachidi, F. Rodriguez, F. Tadeo and J. L. Guzmanb, "Takagi-Sugeno Control of Nocturnal Temperature in Greenhouses Using Air Heating," ISA Transactions, Vol. 50, No. 2, 2011, pp. 315-320.

3. R. F. Escobar, et al., "Sensor Fault Detection and Isolation via High-Gain Observers: Application to a DoublePipe Heat Exchanger," ISA Transactions, Vol. 50, No. 3, 2011; pp. 480-486.

4. M. F. Rahmat, N. A. Mohd Subha, K. M. Ishaq and N. Abdul Wahab, "Modeling and Controller Design for the VVS-400 Pilot Scale Heating and Ventillation System," International Journal on Smart Sensing and Intelligent Systems, Vol. 2, No. 4, 2009, pp. 579-601.

5. H. L. Ho, A. B. Rad, C. C. Chan and Y. K. Wong, "Comparative Studies of Three Adaptive Controllers," ISA Transactions, Vol. 38, No. 1, 1999, pp. 43-53.

6. T. Kealy and A. O'Dwyer, "Closed Loop Identification of a First Order plus Dead Time Process Model under PI Control," Proceedings of the Irish Signals and Systems Conference, University College, Cork, 25-26 June 2002, pp. 9-14.

7. D. M. de la Pena, D. R. Ramirez, E. F. Camacho and T. Alamo, "Application of an Explicit Min-Max MPC to a Scaled Laboratory Process," Control Engineering Practice, Vol. 13, No. 12, 2005, pp. 1463-1471.

8. R. Mooney and A. O'Dwyer, "A Case Study in Modeling and Process Control: The Control of a Pilot Scale Heating and Ventilation System," Proceedings of the 23rd International Manufacturing Conference, University of Ulster, Jordanstown, August 2006, pp. 123-130.

9. N. A. M. Subha, M. F. Rahmat and K. M. Ishaq, "Controller Design for a Pilot-Scale Heating and Ventilation System Using Fuzzy Logic Approach," Jurnal Teknologi Keluaran Khas, Vol. 54, 2011, pp. 123-139.

10. L. P. Wang, "Model Predictive Control System Design and Implementation Using MATLAB," Springer, Berlin, 2009.

11. L. Wang and P. C. Young, "An Improved Structure for Model Predictive Control Using Non-Minimal State Space Realisation," Journal of Process Control, Vol. 16, No. 4, 2006, pp. 355-371.

12. J. M. Maciejowski, "Predictive Control with Constraints," Prentice Hall, Upper Saddle River, 2002.

13. A. Bemporad, F. Borrelli and M. Morari, "Model Predictive Control Based on Linear Programming the Explicit Solution," IEEE Transactions on Automatic Control, Vol. 47, No. 12, 2002, pp. 1974-1985.

14. J. H. Lee and B. L. Cooley, "Min-Max Predictive Control Techniques for a Linear State-Space System with a Bounded Set of Input Matrices," Automatica, Vol. 36, No. 3, 2000, pp. 463-473.

15. S. J. Qin, V. M. Martinez and B. A. Foss, "An Interpolating Model Predictive Control Strategy with Application to a Waste Treatment Plant," Computers and Chemical Engineering, Vol. 21, No. 1, 1997, pp. S881-S886.

16. T. Kawabe, "Robust MPC Method for BMI Based Wheelchair," Intelligent Control and Automation, Vol. 2, No. 2, 2011, pp. 340-350.

17. P. V. Overschee and B. D. Moor, "N4sid: Subspace Algorithms for the Identification of Combined Deterministic-Stochastic Systems," Automatica, Vol. 30, No. 1, 1994, pp. 75-93.

18. M. Verhagen, "Identification of the Deterministic Part of Mimo State Space Models Given in Innovations form from Input-Output Data," Automatica, Vol. 30, No. 1, 1994, pp. 61-74.

19. S. J. Qina, W. Lina and L. Ljung, "A Novel Subspace Identification Approach with Enforced Causal Models," Automatica, Vol. 41, No. 12, 2005, pp. 2043-2053.

20. M. Viberg, "Subspace-Based Methods for the Identification of Linear Time-Invariant Systems," Automatica, Vol. 31, No. 12, 1995, pp. 1835-1852.

21. M. Lovera, T. Gustafsson and M. Verhagen, "Recursive Subspace Identification of Linear and Nonlinear Wiener State Space Models," Automatica, Vol. 36, No. 11, 2000, pp. 1639-1650.

22. T. C. S. Wibowo and N. Saad, "MIMO Model of an Interacting Series Process for Robust MPC via System Identification," ISA Transactions, Vol. 49, No. 3, 2010, pp. 335-347.

23. http://www.didalabdidactique.fr/2008/achat/produit_details.php?id=32&lng=FR

24. E. Yesil, M. Guzelkaya, I. Eksin and O. A. Tekin, "Online Tuning of Set-Point Regulator with a Blending Mechanism Using PI Controller," Turkish Journal of Electrical Engineering, Vol. 16, No. 2, 2008.

25. P. J. Gawthrop and L. Wang, "Intermittent Predictive Control of an Inverted Pendulum," Control Engineering Practice, Vol. 14, No. 11, 2006, pp. 1347-1356.

CHAPTER 11

An Electrothermal Model Based Adaptive Control of Resistance Spot Welding Process

Ziyad Kas, Manohar Das

Department of Electrical and Computer Engineering, Oakland University, Rochester, USA

ABSTRACT

Resistance Spot Welding (RSW) is a process commonly used for joining a stack of two or three metal sheets at desired spots. The weld is accomplished by holding the metallic workpieces together by applying pressure through the tips of a pair of electrodes and then passing a strong electric current for a short duration. Inconsistent weld and insufficient nugget size are some of the common problems associated with RSW. To overcome these problems, a new adaptive control scheme is proposed in this paper. It is based on an electrothermal dynamical model of the RSW process, and utilizes the principle of adaptive one-step-ahead control. It is basically a tracking controller that adjusts the weld current continuously to make sure that the temperature of the workpieces or the weld nugget tracks a desired reference temperature profile. The proposed control scheme is expected to reduce energy consumption by 5% or more per weld, which can result in significant energy savings for any application requiring a high volume of spot welds. The design steps are discussed in details. Also, results of some simulation studies are presented.

Keywords: Resistance Spot Welding, Adaptive Control, Nugget Formation, Energy Saving

1. INTRODUCTION

In resistance spot welding, the welding process begins by applying pressure on a stack of metal sheets, held together between a pair of electrodes. A weld current is then passed through the electrodes, causing resistive heating of the metal workpieces and the formation of a welded joint or nugget, as shown in Figure 1. The formation of a weld nugget strongly depends on the electrical and thermal properties of the sheet and coating materials [1]. Since the contact resistance near the faying surface is much higher than the resistance of the sheets and electrodes, most of the heating is concentrated near the faying surface, causing melting and formation of a nugget there. Depending on the thickness and type of material, welding current ranges from 1,000 to 20,000 amperes or more, while the voltage typically is between 1 and 30 volts [2].

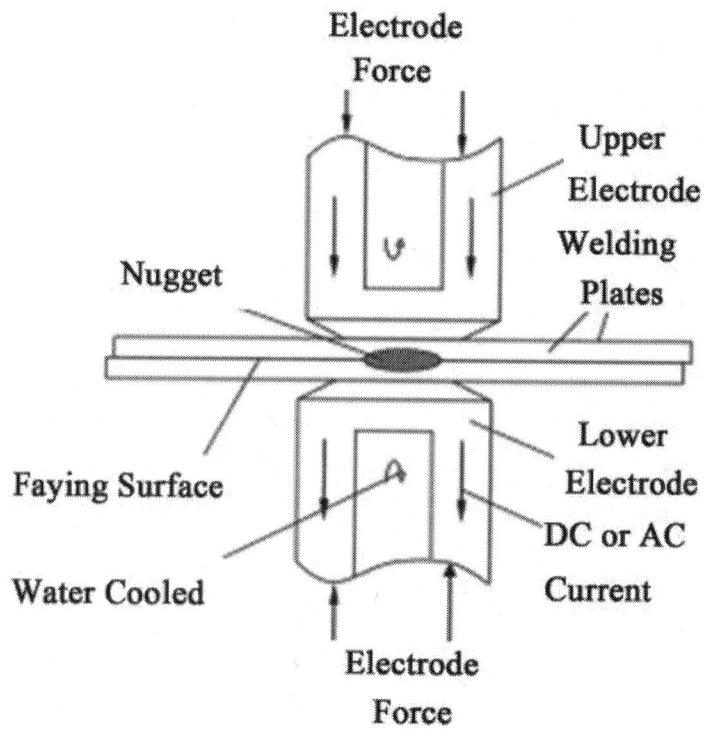

Figure 1. Resistance spot welding system.

A Resistance Spot Welding cycle consists of three main stages as follows:

Stage 1: Squeeze time, which is the time when electrodes press the welded workpieces together.

Stage 2: Weld time, which is the time when welding current is applied producing heat at the faying surface of the workpieces and thus creating a weld nugget.

Stage 3: Hold time, which is the time when electrode force still presses the workpieces together and cools the weld down after the welding current is switched off.

One of the most common applications of resistance spot welding is in the automobile manufacturing industry, where it is used almost universally to weld the sheet metals to form the car body and parts. A typical automotive vehicle today requires about 4000 - 6000 spot welds per vehicle. Considering a worldwide annual production volume of 80 million automotive vehicles, an energy saving RSW controller can result in significant energy savings and reduce carbon footprint accordingly.

During the past two decades, a number of studies have been carried out to improve the RSW process, which focuses on monitoring and control of weld parameters to improve weld quality. The RSW control techniques proposed to date include Proportional-Integral (PI) [3] , Proportional-Derivative (PD) [4] , Proportional-Integral- Derivative (PID) [5] , Fuzzy [6] - [8] , Neural Networks (NN) [9] [10] , or a combination of Fuzzy and NN [11] . The main drawback of these techniques is that they do not take into account the thermal dynamics of the RSW process, i.e. they do not utilize dynamical models that govern the heat transfer and nugget formation in the RSW process. Also, these systems don't take into account any welding process variations, such as variations in coating materials, electrode degradation, and weld force variations.

In this paper, a novel approach to RSW control is presented. This approach has not been explored by other researchers. We start with a simplified heat balance model of a RSW process proposed in [12] and [13], and then use it to design a controller. This thermal model of the heat balance is a function of nugget growth and it determines the temperature variation during welding time. This model is used later to design an adaptive-one-step- ahead (AOSA) controller and an adaptive-weighted one-step-ahead (AWOSA) controller that compensate for unknown process variations and track a desired reference temperature profile. Finally, some simulation results that show the performance of the proposed controllers are presented and compared to the performance of a PID controller. Simulation results show that AOSA and AWOSA controllers are capable of tracking a reference temperature profile when the weld parameters are unknown, as well as reduce the energy needed to make a weld by 6%.

The organization of this paper is as follows. Section 2 presents a simplified electrothermal dynamical model of a RSW nugget formation process. The design of adaptive OSA and WOSA controllers is discussed in Section 3. Section 4 presents the results of some simulation studies, and finally some concluding results are provided in Section 5.

2. ELECTROTHERMAL DYNAMICAL MODEL OF A RSW NUGGET FORMATION PROCESS

To start with, we consider a simplified heat balance model of a RSW process, presented in [13] . The simplified dynamical model of a RSW process determines the heat balance in the system as a function of nugget temperature. For a simplified nugget model, shown in Figure 2, the heat balance can be described by the following equations:

The total heat generation rate, $\dot{Q}_g(t)$ is given by

$$\dot{Q}_g(t) = I^2(t)R(t) \tag{1a}$$

$$R(t) = R_w + R_c + R_e \tag{1b}$$

where $I(t)$ denotes the welding current, and $R(t)$ denotes the total resistance consisting of the resistance of work pieces, R_w, contact resistance, R_c, and electrode resistance, R_e. Since R_w and R_e are very small compared to the total contact resistance R_c, R_w and R_e can be neglected in (1b).

The total contact resistance can then be described as,

$$R_c = R(t)_{\text{electrode-sheet}} + R(t)_{\text{sheet-sheet (faying surface)}} \tag{1c}$$

A linear relationship between the resistance and temperature is assumed to model the heat generated as a function of temperature. Thus,

$$R(t)_{\text{electrode-sheet}} = \rho \frac{2l_1}{A} \tag{1d}$$

$$R(t)_{\text{sheet-sheet(faying surface)}} = \rho \frac{2p}{A} \tag{1e}$$

$$\rho = \rho(T) = \rho_o \left[1 + \alpha_r \left(\theta - \theta_o\right)\right] \tag{1f}$$

where ρ denotes the resistivity of the material, l_1 denotes the distance from the melting interface to electrode contact surface, p denotes the penetration, A is the cross sectional area, ρ_o denotes the resistivity at reference temperature θ_o, θ and α_r are the temperature to be controlled and the temperature coefficient respectively.

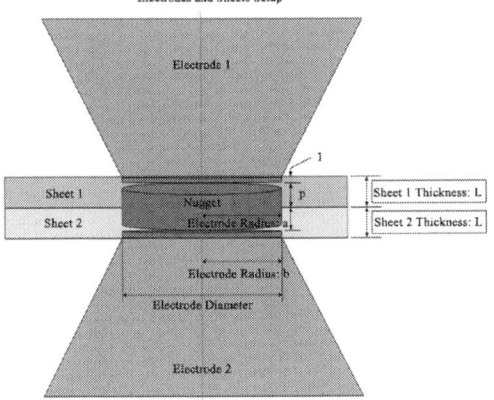

Figure 2. A simplified model of a weld nugget.

Substituting (1f) in (1d) and (1e) we get

$$R(t)_{\text{electrode-sheet}} = c_1\theta(t) + c_2$$

(1g)

$$R(t)_{\text{sheet-sheet(faying surface)}} = c_3\theta(t) + c_4$$

(1h)

where

$$c_1 = \frac{2l_1\rho_\circ\alpha}{A}$$

(1i)

$$c_2 = \frac{2l_1\rho_\circ}{A}(1 - \alpha_r\theta_\circ)$$

(1j)

$$c_3 = \frac{2p\rho_\circ\alpha}{A}$$

(1k)

$$c_4 = \frac{2p\rho_\circ}{A}(1 - \alpha_r\theta_\circ)$$

(1l)

Substituting (1g) and (1h) in (1a) we get

$$\dot{Q}_g(t) = I^2(t)\left(c_1\theta(t) + c_2 + c_3\theta(t) + c_4\right)$$

(1m)

$$= c_5 I^2(t)\theta(t) + c_6 I^2(t)$$

(1n)

where

$$c_5 = c_1 + c_3$$

(1o)

$$c_6 = c_2 + c_4$$

(1p)

The heat of fusion required for nugget formation is given by:

$$H_f = H\Delta V_n$$

(2a)

$$\Delta V_n = \pi a^2 p \tag{2b}$$

where H denotes the heat of fusion per unit volume, ΔV_n denotes the nugget volume, and p, a denote the penetration and nugget radius respectively.

Substituting (2b) in (2a) and normalizing over the weld duration, Δt, we get the heat of fusion per unit time:

$$\frac{H_f}{\Delta t} = H\pi a^2 p = c_7 \tag{2c}$$

Neglecting the heat loss in the surroundings and the electrodes, the heat required to raise temperature by $d\theta(t)$ is given by

$$dQ_T(t) = \rho C_p d\theta(t)\Delta V \tag{3a}$$

where ρ denotes the density, C_p denotes the specific heat, V is the volume, and $d\theta(t)$ is the temperature rise. We rewrite (3a) as:

$$dQ_T(t) = c_8 d\theta(t) \tag{3b}$$

where

$$c_8 = \rho C_p \pi a^2 p \tag{3c}$$

The total heat loss rate is given by

$$\dot{Q}_L(t) = \dot{Q}_a(t) + \dot{Q}_r(t) \tag{4a}$$

$$= k_1 \pi a^2 \left[\frac{\theta(t)-\theta_1}{l_1} + \frac{10\theta(t)\beta L}{b\sqrt{\alpha}} \right]$$

$$= \left(\frac{k_1 \pi a^2}{l_1} + \frac{10 k_1 \pi a^2 \beta L}{b\sqrt{\alpha}} \right)\theta(t) - \frac{k_1 \pi a^2 \theta_1}{l_1}$$

$$= c_9 \theta(t) - c_{10} \tag{4b}$$

where

$$c_9 = \left(\frac{k_1 \pi a^2}{l_1} + \frac{10 k_1 \pi a^2 \beta L}{b\sqrt{\alpha}} \right) \tag{4c}$$

$$c_{10} = \frac{k_1 \pi a^2 \theta_1}{l_1} \tag{4d}$$

In the above equations, $\dot{Q}_a(t)$ and $\dot{Q}_r(t)$ denote the axial and radial loss rates, respectively; k_1 repre-

sents thermal conductivity, a is the nugget radius; $\theta(t)$, θ_1, represent the melting temperature and the interface temperature at the work piece respectively; l is the distance from the melting interface to the electrodes contact area; β represents the final penetration to work piece thickness ratio; L is the sheet thickness; b, α represent the electrode radius and thermal diffusivity of work piece respectively.

The heat balance equation over time $(t, t+dt)$ is given by

$$\dot{Q}_g(t) = \frac{H_f}{\Delta t}dt + dQ_T(t) + \dot{Q}_L(t)dt \tag{5}$$

Substituting (1n), (2c), (3b), and (4b) in (5) and rearranging it, we get

$$c_8\frac{d\theta(t)}{dt} = c_5 I^2(t)\theta(t) + c_6 I^2(t) - c_9\theta(t) + c_{10} - c_7 \tag{6a}$$

or, equivalently,

$$\frac{d\theta(t)}{dt} = c_{11} I^2(t)\theta(t) + c_{12} I^2(t) - c_{13}\theta(t) + c_{14} \tag{6b}$$

where

$$c_{11} = c_5/c_8 \tag{6c}$$

$$c_{12} = c_6/c_8 \tag{6b}$$

$$c_{13} = c_9/c_8 \tag{6c}$$

$$c_{14} = (c_{10} - c_7)/c_8 \tag{6d}$$

For the sake of notational convenience, let $y(t) = \theta(t)$ and $u(t) = I^2(t)$. Then (6b) can rewritten as

$$\frac{dy(t)}{dt} = c_{11}u(t)y(t) + c_{12}u(t) - c_{13}y(t) + c_{14} \tag{7}$$

Equation (7) represents a bilinear electrothermal dynamical model of a RSW process. Note that this simplified model neglects the heat required to raise the temperature of the electrodes and the nugget surroundings. Also, it assumes that most of the heating occurs near the faying surface due to its high contact resistance. The size of the workpieces is assumed to be infinite in the radial direction and the nugget shape is assumed to be a disk growing radially and axially in the same proportions. The nominal nugget diameter is assumed to be $4.5\sqrt{L}$, where L is the sheet thickness.

Using a first order Euler approximation for $\frac{dy}{dt}$ with a sampling period T_s, the following discrete time equation is derived from the system Equation (7):

$$\frac{y(k+1)-y(k)}{T_s} = c_{11}u(k)y(k)+c_{12}u(k)-c_{13}y(k)+c_{14}$$

(8a)

or

$$y(k+1) = Ay(k)+Bu(k)+Cu(k)y(k)+D$$

(8b)

where

$$A = 1 - c_{13}T_s$$

(8c)

$$B = c_{12}T_s$$

(8d)

$$C = c_{11}T_s$$

(8e)

$$D = c_{14}T_s$$

(8f)

Also, k denotes the discrete time index $(k=0,1,2,\cdots)$ and kT_s denote the sampling instances. The above electrothermal model is characterized by four unknown parameters, namely, A, B, C, and D.

3. DESIGN OF A RSW CONTROLLER

To develop a control scheme for controlling the nugget temperature of the RSW model presented by Equation (8a), we realize that it presents a bilinear system characterized by some unknown parameters. These parameters can vary from weld to weld, and in most cases we have no prior knowledge of the parameter values. In view of this, we propose to use an adaptive OSA and WOSA controllers.

The proposed adaptive control scheme involves measurement of the inputs and outputs of the system, estimation of unknown system parameters using a recursive least squares (RLS) parameter estimation algorithm, and computation of a control signal based on the estimated parameter values. Also, the temperature of the weld nugget is monitored indirectly by assuming it to be proportional to the contact resistance.

3.1. Adaptive OSA and WOSA Controllers

In an adaptive controller, the sampled measurements, $u(k)$ and $y(k)$, are used to estimate the model parameters, A, B, C and D in Equation (8b), using a recursive parameter estimation method, such as recursive least square (RLS). The estimated values of these parameters are then used to compute the OSA/WOSA control signals.

3.2. Parameter Estimation

First we write model Equation (7) in the following form:

$$y(k+1) = \varphi(k)^{\mathrm{T}} X^*$$

(9a)

where

$$\varphi(k)^{\mathrm{T}} = \begin{bmatrix} y(k-1) & u(k-1) & y(k-1)*u(k-1) & 1 \end{bmatrix}^{\mathrm{T}}$$

(9b)

$$X^* = \begin{bmatrix} A & B & C & D \end{bmatrix}^{\mathrm{T}}$$

(9c)

Next, the estimated value of θ_o is computed recursively using the following RLS algorithm:

$$\hat{\theta}(k) = \hat{\theta}(k-1) + \frac{P(k-2)\varphi(k-1)}{1+\varphi(k-1)^{\mathrm{T}} P(k-2)\kappa(k-1)} \left[y(k) - \varphi(k-1)^{\mathrm{T}} \hat{\theta}(k-1) \right]; \quad k \geq 1$$

(10a)

$$P(k-1) = P(k-2) - \frac{P(k-2)\varphi(k-1)\varphi(k-1)^{\mathrm{T}} P(k-2)}{1+\varphi(k-1)^{\mathrm{T}} P(k-2)\varphi(k-1)}$$

(10b)

$$\hat{X}(0) = \begin{bmatrix} \gamma & 0 & 0 & 0 \end{bmatrix}^{\mathrm{T}}$$

(10c)

$$P(-1) = \sigma I$$

(10d)

where $\gamma > 0$ is a small number and $\sigma > 0$ is chosen to be large. Also, $\hat{c}(k)$ is always constrained to be non-negative, i.e.,

$$\hat{C}(k) > \varepsilon > 0 \quad \text{for all } k$$

(10e)

Given an estimate $\hat{X}(k)$ of X^*, we define the predicted output at time $k+1$ as:

$$\hat{y}(k+1) = \varphi(k)^{\mathrm{T}} \hat{X}(k)$$

(11)

3.3. Adaptive-One-Step-Ahead Tracking Controller

One-step-ahead (OSA) control scheme for linear systems has been well investigated in [14]. An OSA controller attempts to bring the predicted output, $y(k+1)$ at time $k+1$, to the desired value, $y^*(k+1)$ in one step. Thus, it minimizes the following cost function:

$$J_1(k+1) = \frac{1}{2}\left[y(k+1)-y^*(k+1)\right]^2$$

(12)

The corresponding OSA control law is given by [14]:

$$\bar{u}(k) = \frac{y^*(k+1)-Ay(k)-D}{B+Cy(k)}$$

(13)

The above control signal needs to be constrained by the maximum current delivery capacity of the controller, u_{max}, as follows:

$$u(k) = \begin{cases} \bar{u}(k), & \text{if } 0 < \bar{u}(k) < u_{max} \\ 0, & \text{if } \bar{u}(k) \le 0 \\ u_{max}, & \text{if } \bar{u}(k)) \ge u_{max} \end{cases}$$

(14)

The adaptive OSA controller uses the estimate, $\hat{X}(k)$ in Equation (11) to compute the control signal, $u(k)$, from the following adaptive version of Equation (13) above:

$$\bar{u}(k) = \frac{y^*(k+1)-\hat{A}(k)y(k)-\hat{D}(k)}{\hat{B}(k)+\hat{C}(k)y(k)}$$

(15)

where $\hat{A}(k), \hat{B}(k), \hat{C}(k),$ and $\hat{D}(k)$ denote the estimated values of $A, B, C,$ and $D,$ respectively, at time $k.$

One of the potential drawbacks of OSA controllers is excessive control efforts that often result from attempting to bring $y(k+1)$ to $y^*(k+1)$ in one step. To address this potential problem, an AWOSA controller is discussed below.

3.4. Adaptive Weighted One-Step-Ahead Controller

The excessive effort to bring the output $y(k+1)$ to the desired value $y^*(k+1)$ in one step using AOSA may result in an unfavorable saturation of the input. The adaptive weighted one-step-ahead controller attempts to seek a tradeoff between tracking accuracy and control effort by considering a slight generalization of the cost function (12) to the form (16) given below. Thus, it minimizes the following cost function:

$$J_2(k+1) = \frac{1}{2}\left[y(k+1) - y^*(k+1)\right]^2 + \frac{\lambda}{2}u(k)^2$$

(16)

where, $0 < \lambda < 1$ is chosen to provide a desired tradeoff.
The minimization of the cost function in (16) leads to the weighted one-step-ahead control law [14]:

$$\bar{u}(k) = \frac{\left(B + Cy(k)\right)\left(y(k+1) - Ay(k) - D\right)}{\left(B + Cy(k)\right)^2 + \lambda}$$

(17)

The above control law is also constrained by the maximum current delivery capacity, u_{max}, as shown in Equation (14) above. The choice of λ provides a desired tradeoff between tracking accuracy and control effort. A small λ results in good tracking but requires high level of control effort. A large λ, on the other hand, reduces control efforts at the cost of tracking accuracy.

The adaptive WOSA controller uses the estimate, $\hat{X}(k)$, in Equation (11) to compute the control signal, $u(k)$ from the following adaptive version of Equation (17) above:

$$\bar{u}(k) = \frac{\left(\hat{B}(k) + \hat{C}(k)y(k)\right)\left(y(k+1) - \hat{A}(k)y(k) - \hat{D}(k)\right)}{\left(\hat{B}(k) + \hat{C}(k)y(k)\right)^2 + \lambda}$$

(18)

where $\hat{A}(k), \hat{B}(k), \hat{C}(k),$ and $\hat{D}(k)$ denote the estimated values of $A, B, C,$ and $D,$ respectively, at time $k.$

4. SIMULATION RESULTS AND DISCUSSION

This section presents the results of a simulation study showing the performance of the system with the proposed AOSA and AWOSA controllers and also compare them with a PID controller. Each controller is designed for tracking a reference temperature profile.

The reference temperature profile is a good indicator of the weld quality. Therefore, it is desirable to keep the temperature variation close to a desired variation curve, which may be experimentally predetermined for the good welds. A typical reference temperature profile for good weld is shown inFigure 3 below [1] . Basically, such a curve is characterized by a fast rise of temperature to melting point, melting of the workpieces at the faying surface area which causes a slight drop in temperature, followed by a cooling zone that results from removal of weld current. The actual nugget temperature is measured during the weld cycle using the relationship described by Equation (1f). Depending on the tracking error signal, the welding current is adjusted so as to reduce the temperature error.

For these simulations, we have selected two sheets of mild steel with the same thickness as the materials to be welded. The force variation and electrode wear are considered as unknown process variables that impact the nugget size (diameter and penetration). The Figures below show the performance of the AOSA, AWOSA, and PID controllers due to 20% increase in nugget diameter and 50% increase in indentation from their desired values.

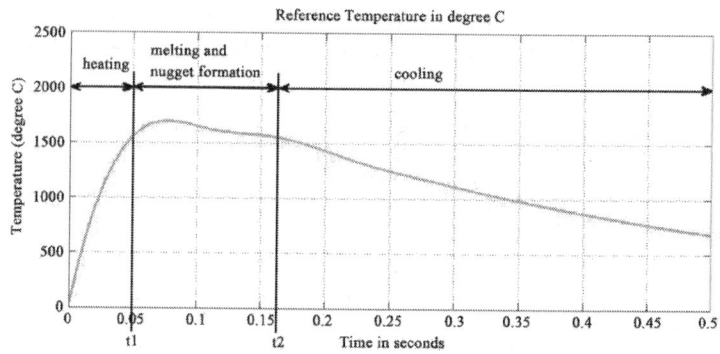

Figure 3. Desired reference temperature profile.

Figure 4 shows the performance of the AOSA controller using $I_{max} = 12\,\text{KA}$, where I_{max} denotes the maximum current delivery capacity of the weld controller. We can see that the AOSA controller adapts to the parameter change and force the output temperature profile to follow the desired temperature profile. Also, we can see that the energy required for the weld is lower than that of the PID controller.

Figure 5 and Figure 6 show the performance of AWOSA controller using $I_{max} = 12\,\text{KA}$ with $\lambda = 0.1$ and

1, respectively. Here we notice that when λ is high, the output temperature profile does not follow the desired output temperature profile well. However, increasing λ results in decreasing the total energy required for the weld.

Figure 7 shows the performance of the PID controller prior to any parameter change using $I_{max} = 12\ KA$. After multiple trial and error attempts to get satisfactory results, the parameters of the PID controllers are: Proportional (P) = 0.5, Integral (I) = 26.56, Derivative (D) = 0.

In Figure 8 we see that the PID controller looses track of the reference temperature profile due to weld parameters change. Also, we can see that PID controller requires more energy for the weld comparing to AOSA and AWOSA.

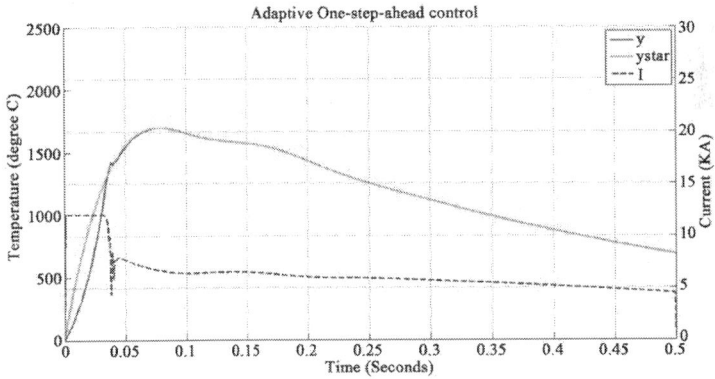

Figure 4. Performance of AOSA Controller with 20% increase in nugget diameter and 50% increase in indentation; Imax = 12 KA Energy=2558 W.

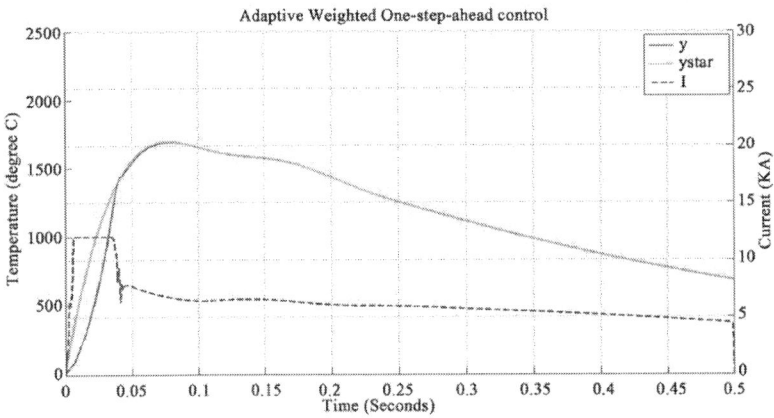

Figure 5. Performance of AWOSA Controller with 20% increase in nugget diameter and 50% increase in indentation; = 0.1, Imax = 12 KA Energy=2558 W.

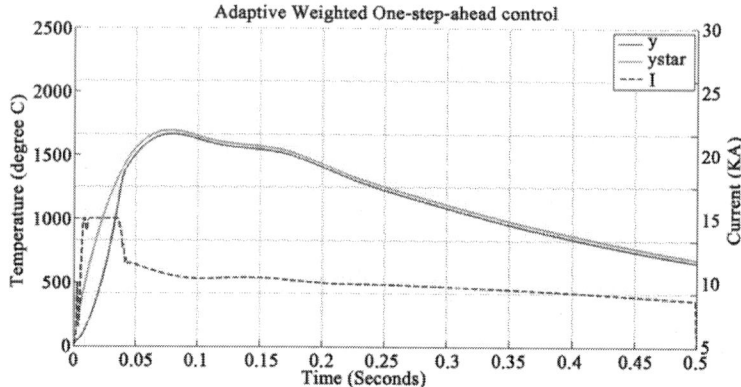

Figure 6. Performance of AWOSA Controller with 20% increase in nugget diameter and 50% increase in indentation; =1, Imax = 12 KA Energy=2470 W.

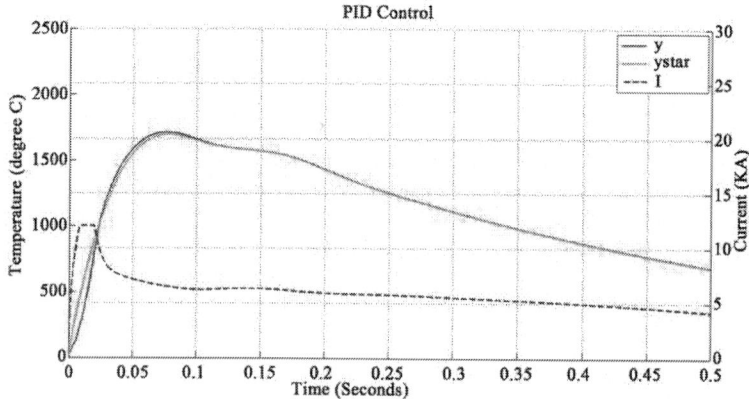

Figure 7. Performance of PID Controller prior to unknown parameter variations; Imax = Energy=2393 W 12 KA,.

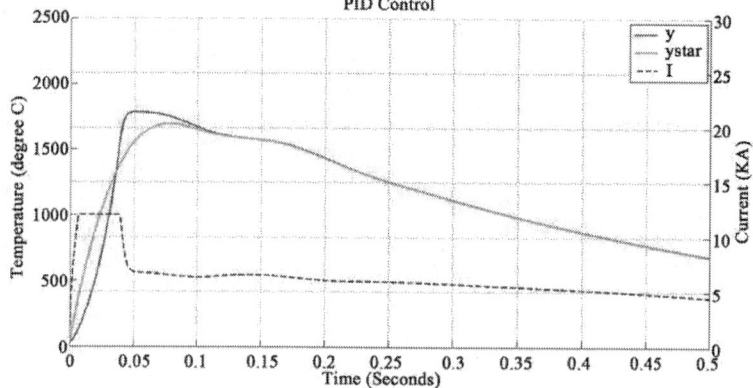

Figure 8. Performance of PID Controller with 20% increase in nugget diameter and 50% increase in indentation; Imax =12 KA, Energy 2632 W.

Comparing the simulation results for the three controllers, we can see that AOSA and AWOSA controllers compensate for the parameter variations and track the reference temperature profile quite well. Simulation results in Figure 5 for the AWOSA controller show satisfactory performance and a good tradeoff between tracking error and total energy required for the weld regardless of change in weld parameters. The output temperature profile follows the desired temperature profile reasonably well during the heating stage prior to the melting point. Also, we can see that the total energy required to make a weld using AWOSA is reduced by 6% comparing to the PID controller when $I_{max} = 12\,\mathrm{KA}$. This can result in significant energy savings for applications requiring a high volume of spot welds, such as manufacturing of automotive vehicles.

5. CONCLUSION

This paper presents a new approach for designing adaptive OSA and WOSA controllers for resistance spot welding processes by utilizing a simplified electrothermal dynamical model of the process. Simulation results of AOSA and AWOSA performance are compared with those of a PID controller. These results indicate that using the proposed AOSA and AWOSA controllers, the nugget temperature profile is forced to track a desired reference temperature profile in presence of unknown parameter variations. Also, these controllers reduce the energy consumed to perform a spot weld, which can result in significant energy savings for applications requiring a high volume of spot welds, such as manufacturing of automotive vehicles.

REFERENCES

1. Zhang, H. and Senkara, J. (2012) Resistance Welding Fundamentals and Applications. Taylor & Francis Group, Boca Raton.
2. Govik, A. (2009) Modeling of the Resistance Spot Welding Process. M.S. Thesis, Institute of Technology, Linkopings University, Linkoping.
3. Won, Y.J., Cho, H.S. and Lee, C.W. (1983) A Microprocessor-Based Control System for Resistance Spot Welding Process. Proceedings of ACC, San Francisco, 22-24 June 1983, 734-738.
4. Zhou, K. and Cai, L. (2014) A Nonlinear Current Control Method for Resistance Spot Welding. Proceedings of ASME Transactions on Mechatronics, 19, 559-569.
5. Salem, M. and Brown, L.J. (2011) Improved Consistency of Resistance Spot Welding with Tip Voltage Control. Proceedings of CCECE, Niagara Falls, 8-11 May 2011, 548-551.

258 Advanced Model Predictive Control

6. Chen, X., Araki, K. and Mizuno, T. (1997) Modeling and Fuzzy Control of the Resistance Spot Welding Process. Proceedings of SICE, Tokushima, 29-31 July 1997, 898-994.

7. El-Banna, M., Filev, D. and Chinnam, R.B. (2006) Intelligent Constant Current Control for Resistance Spot Welding. Proceedings of IEEE Conference on Fuzzy Systems, Vancouver, 16-21 July 2006, 1570-1577.

8. Chen, X. and Araki, K. (1997) Fuzzy Adaptive Process Control of Resistance Spot Welding with a Current Reference Model. Proceedings of IEEE Conference on Intelligent Processing Systems, Beijing, 28-31 October 1997, 190-194.

9. Shriver, J., Peng, H. and Hu, S.J. (1999) Control of Resistance Spot Welding. Proceedings of ACC, San Diego, 2-4 June 1999, 187-191.

10. Ivezic, N., Allen Jr, J.D. and Zacharia, T. (1999) Neural Network-Based Resistance Spot Welding Control and Quality Prediction. Proceedings of IPMM, Honolulu, 10-15 July 1999, 989-994.

11. Messler Jr, R.W., Jou, M. and Li, C.J. (1995) An Intelligent Control System for Resistance Spot Welding Using a Neural Network and Fuzzy Logic. Proceeding of IAC, Orlando, October 1995, 1757-1763.

12. Kim, E.W. and Eagar, T.W. (1988) Parametric Analysis of Resistance Spot Welding Lobe Curve. SAE Technical Paper Series, Warrendale.

13. Kas, Z. and Das, M. (2014) A Thermal Dynamical Model Based Control of Resistance Spot Welding. Proceedings of IEEE EIT 2014, Milwaukee, 5-7 June 2014, 264-269.

14. Goodwin, G.C. and Sin, K.S. (1983) Adaptive Filtering Prediction and Control. Prentice-Hall, Englewood Cliffs.

CHAPTER 12

Nonlinear Predictive Control of a Hydropower System Model

Runfan Zhang, Diyi Chen and Xiaoyi Ma*

Institute of Water Resources and Hydropower Research, Northwest A&F University, Yangling 712100, China

ABSTRACT

A six-dimensional nonlinear hydropower system controlled by a nonlinear predictive control method is presented in this paper. In terms of the nonlinear predictive control method; the performance index with terminal penalty function is selected. A simple method to find an appropriate terminal penalty function is introduced and its effectiveness is proved. The input-to-state-stability of the controlled system is proved by using the Lyapunov function. Subsequently a six-dimensional model of the hydropower system is presented in the paper. Different with other hydropower system models; the above model includes the hydro-turbine system; the penstock system; the generator system; and the hydraulic servo system accurately describing the operational process of a hydropower plant. Furthermore, the numerical experiments show that the six-dimensional nonlinear hydropower system controlled by the method is stable. In addition, the numerical experiment also illustrates that the nonlinear predictive control method enjoys great advantages over a traditional control method in nonlinear systems. Finally, a strategy to combine the nonlinear predictive control method with other methods is proposed to further facilitate the application of the nonlinear predictive control method into practice.

Keywords: nonlinear predictive control; hydropower system; Lyapunov function; control method; stability

1. INTRODUCTION

Hydropower, as a low-cost, zero-pollution and renewable energy, has been developed since the twentieth century [1]. Thanks to modern technologies, hydro-turbines are becoming bigger and bigger. As a result, many hydropower plants with great capacity are being built around the world to generate electricity to resolve the serious energy problem. The system of such a powerful hydropower plant including penstock systems, water turbines, generators, regulators and loads, is so complex that it is difficult to control [2]. However, the stability of a hydropower system plays an important role in the stability of the whole power system and the plants themselves [3]. Thus, the issue of the control and stability of hydropower systems enjoys great popularity among researchers. For example, in [4], a new adjustment method of PID governors was proposed for hydropower plants with long penstocks to control the power frequency. A nonlinear hydropower plant system was controlled by a fuzzy control method in [2]. A micro-hydro power plant model with a smaller, lighter, more robust and more efficient higher-speed turbine was built in [5], and a control scheme was also proposed. In [6], an accurate new dynamical model for a cascaded hydropower plant, a complex nonlinear system involving interacting input and output nonlinear parameters, nonlinear flow rates, and nonlinear dynamical hydraulic heads was developed. The primary control system and stability analysis of a hydropower plant was discussed in [7], and controllers, including PI, PID, and PI-PD, were studied. In [8], a control method based on integrating the entropy and mean value of the tracking error with the constraints was proposed for hydro-turbine speed governors.

The control problem relates to making the system reach the expect points or states. At present, there are an increasing number of control theories, and parts of them even have been applied in practice for many years such as PID control [8,9], adaptive control [10,11,12,13], and feedback control [14,15,16] and so on. On the contrary, many remarkable state-of-the-art control methods, which enjoy a variety of advantages compared to traditional control methods, are being developed. For instance, a discrete sliding mode controller coupled with a Kalman filter was designed in [17,18,19] to control the combustion phasing in real-time model based on the control method of Homogenous Charge Compression Ignition. However, the sliding model control has the chattering problem when the controllers switch to the other states, so a no-chattering sliding model control method was proposed in [20]. A kind of switched uncertain nonlinear system was controlled by an adaptive fuzzy tracking control method in [21]. In [22], multi-zone buildings were identified by an artificial neural network and controlled by predictive control techniques. The robust $H\infty$ finite-time control method for a discrete system was discussed in [23]. Compared with the abovementioned methods, nonlinear predictive control theory is easier to apply in practice, and can be combined with traditional methods, which takes advantages of both traditional methods and predictive control as shown later in this paper.

Predictive control is an optimization problem. Using the discrete model of a system, the system's future states with the controllers are predicted. Then, based on the predicted values, appropriate controllers are found to minimize the performance index [24]. This control method is called Model Predictive Control (MPC), and if the controlled system is a nonlinear system, it will be Nonlinear Model Predictive Control (NMPC). The predicted distance is called the predictive length. In fact, only the first set of predicted controllers is selected to control the system to obtain the states of the next step. This relates to the reduction of the errors [24,25]. In the performance index, the terminal penalty function, the distance between states and the expected states, and the distance between the controllers and the expected controllers, which generally are zeros, are considered [24]. The concept of predictive control was originated in 1967 by Lee and Markus [24,25]. However, not only the hardware but also the software couldn't achieve very rapidly their computing mission in Lee's and Markus' paper. Thus, predictive control was only a concept in that time. Thanks to modern technologies, predictive control started enjoying popularity in the late 1970s. Later, in 1982 Chen and Shaw first brought the concept of nonlinear predictive control in continuous time proved by the Lyapunov function [26]. Since then the predictive control method has aroused attention. For example, many remarkable studies have been done in different fields [27,28,29,30,31,32,33,34], such as large-scale optimization problems [29], hybrid electric vehicles [30], and the shell heavy oil fractionator benchmark control problem [31], and so on.

Motivated by what has been discussed above, we provide a simple method to select the performance index for a discrete time model proved by the Lyapunov function. The proof process is so intelligible that experts and technicians can understand and apply this method without too much effort. A six-dimensional complex nonlinear hydropower plant system is controlled based on the proposed method. In addition, we supply instructions for technicians in other fields or other hydropower plants to apply this method. Moreover, the strategy is also introduced to combine the NMPC with other traditional control methods.

The remaining parts of this paper are organized as follows: in Section 2, the principal concepts including a simply method to select performance index and the effectiveness of the proposed method are presented. A nonlinear hydropower plant model is introduced in Section 3. The main results of the controlled hydropower system are illustrated in Section 4. In Section 5, we discuss the effectiveness of the NMPC and how to apply this method in the other situations, involving the combination of the NMPC with other control methods.

2. NONLINEAR PREDICTIVE CONTROL

2.1. System Model

The continuous system can be described as:

$$\dot{\mathbf{x}} = \mathbf{F}(\mathbf{x}, \mathbf{u}),\tag{1}$$

where $\mathbf{x} \in R^n$, $\mathbf{u} \in R^m$, the constraints of x and u are $\mathbf{x} \in X$, $\mathbf{u} \in U$, respectively; and f($\cdot$) is a function in $X \times U$ and Lipschitz continuous.

Assume that Equation (1) has an equilibrium point $(x_0, u_0) = (0,0)$. If $F(\cdot)$ does not follow the assumption, we should move the equilibrium point (x0,u0) to (0,0) through transformation of coordinates. Equation (1) can be discretized as:

$$x_{k+1} = f\left(x_k, u_k\right),\qquad(2)$$

where f(·) is a function in X×U, and is also Lipschitz continuous.

Note that if the system is a discrete system, Equation (2) is able to directly describe the system.

2.2. Nonlinear Predictive Control Theory

The concept of predictive control is that the system can be predicted by Equation (2). Based on the state vector xk with time k, several future state vectors of the system are able to be predicted with the input parameters uk as:

$$x_{k+i+1} = f\left(x_{k+i}, \bar{u}_{k+i}\right),\left(i = 0,\cdots,N\right),\qquad(3)$$

where x_{k+} are the predicted values based on the states x_k and $\bar{u}_{k+i}$ are the predictive controllers based on the states xk calculated later, N is the predict length.

The controllers $\bar{u}_{k+i}$ are supposed to make the performance index Jk be minimum. The Jk is always defined as:

$$J_k = \sum_{j=0}^{N-1} L\left(x_{k+j}, \bar{u}_{k+j}\right) + V\left(x_{k+N}\right),\qquad(4)$$

In this case, the predictive control issue is to solve the following optimization problem:

$$\bar{u}_{k+j} \mid J_k = \sum_{\substack{min \\ j=0}}^{N-1} L\left(x_{k+j}, \bar{u}_{k+j}\right) + V\left(x_{k+N}\right),\quad x \in X,\quad u \in U,\quad \text{and}\quad x_{k+N} \in \Omega_\alpha,$$

$$(5)$$

where L(·) should follow Assumption 1, V(·) is the terminal penalty function subjected to Assumption 2, and Ωα is the terminal region defined later. However, after the sets of controllersu̅k+i are calculated, in order to reduce the errors caused by the controllers, only the first set controllers:

$$\mathbf{u}_k^* = \bar{\mathbf{u}}_k \tag{6}$$

are working in one step at the system, which is:

$$\mathbf{x}_{k+1} = \mathbf{f}\left(\mathbf{x}_k, \mathbf{u}_k^*\right). \tag{7}$$

Assumption 1
(1). L(0,0)=0;
(2). L(·) is Lipschitz continuous;
(3). There exists a positive parameter δ satisfying L(x,u)≥δ‖(x,u)‖.
Assumption 2
There exist a local controller uv satisfying
(1). V(f(x,uv))−V(x)≤−L(x,uv);
(2). V(·) is Lipschitz continuous.
In order to determine an appropriate terminal penalty function V(·) to make the system be stable or asymptotic stable, the following efforts are needed. We linearize the system (1) by Jacobian method at the origin as:

$$\dot{\mathbf{x}} = A\mathbf{x} + B\mathbf{u}, \tag{8}$$

where $A = \left.\dfrac{\partial \mathbf{F}}{\partial \mathbf{x}}\right|_{(x,u)=(0,0)}$ and $B = \left.\dfrac{\partial \mathbf{F}}{\partial \mathbf{u}}\right|_{(x,u)=(0,0)}$. If the system (8) can be stabilized, the linear feedback controllers uf=Kx can be found to stabilize the system (7). Then the system is:

$$\dot{\mathbf{x}} = G\mathbf{x} \tag{9}$$

where G=A+BK.

Theorem 1

If the function L(·) is defined as $L(\mathbf{x},\mathbf{u}) = \mathbf{x}^T Q\mathbf{x} + \mathbf{u}^T R\mathbf{u}$, where Q and R are positive symmetric matrixes, respectively. The terminal penalty function $V(\cdot)$ will be $V(\mathbf{x}) = \mathbf{x}^T P\mathbf{x}$, where P is a positive-definite matrix solved by the Lyapunov equation as:

$$(G+\kappa I)^T P + P(G+\kappa I) = -(Q + K^T RK), \quad \kappa < -\lambda_{max}(G) \tag{10}$$

Proof

Defining $Q^* = Q + K^T RK$, the Lyapunov equation is:

$$(G + \kappa I)^T P + P(G + \kappa I) = -Q^* . \tag{11}$$

It is easy to know that Q* is positive-definite and symmetric. As mentioned above, if the system (9) is asymptotic stable, the real parts of all eigenvalues of G will be negative, *i.e.*,

$$\mathbf{Re}(\lambda(G)) < 0 \tag{12}$$

Since $\kappa < -\lambda_{max}(G)$, it is simple to know that the real parts of all eigenvalues of $(G + \kappa I)$ are negative. Thus, in accordance with the solvability condition of Lyapunov equation, P solved by Equation (11) is positive-definite and symmetric.

For a discrete system, the differential can be defined as the difference value between two steps for a per unit length. Thus:

$$\frac{d}{dt} \mathbf{x}^T P \mathbf{x} = V(\mathbf{x}_{k+1}) - V(\mathbf{x}_k) = V(f(\mathbf{x}, \mathbf{u}_v)) - V(\mathbf{x}) \tag{13}$$

where $\mathbf{u}_v = K\mathbf{x}$ are feedback controllers. Substituting uv for u in the system (1), we have:

$$\dot{\mathbf{x}} = \mathbf{F}(\mathbf{x}, \mathbf{u}_v) = \mathbf{F}(\mathbf{x}, K\mathbf{x}) . \tag{14}$$

Then:

$$\begin{aligned}
\frac{d}{dt} \mathbf{x}^T P \mathbf{x} &= \left(\frac{d\mathbf{x}}{dt}\right)^T P \mathbf{x} + \mathbf{x}^T P \left(\frac{d\mathbf{x}}{dt}\right) \\
&= (G\mathbf{x} + \varphi(\mathbf{x}))^T P \mathbf{x} + \mathbf{x}^T P (G\mathbf{x} + \varphi(\mathbf{x})) \\
&= \mathbf{x}^T G^T P \mathbf{x} + \mathbf{x}^T P G \mathbf{x} + \varphi^T(\mathbf{x}) P \mathbf{x} + \mathbf{x}^T P \varphi(\mathbf{x}) \\
&= \mathbf{x}^T (G^T P + PG) \mathbf{x} + 2\mathbf{x}^T P \varphi(\mathbf{x})
\end{aligned} \tag{15}$$

where $\varphi(\mathbf{x}) = \mathbf{F}(\mathbf{x}, K\mathbf{x}) - G\mathbf{x}$.

The function $\mathbf{x}^T P \varphi(\mathbf{x})$ is bounded and its bound is:

$$\mathbf{x}^T P \varphi(\mathbf{x}) \leq \|\mathbf{x}^T P\| \cdot \|\varphi(\mathbf{x})\| \leq \|P\| \cdot L_\varphi \|\mathbf{x}\|_P^2 \leq \frac{\|p\| \cdot L_\varphi}{\lambda_{min}(P)} \|\mathbf{x}\|_P^2 , \tag{16}$$

where $L_{\varphi}=\sup\left\{\frac{\|\varphi(x)\|}{\|x\|}\mid x\in\Omega_{\alpha_1},x\neq0\right\}$, and selecting an appropriate α satisfying $0<\alpha\leq\alpha_1$ enables:

$$L_{\varphi}\leq\frac{\kappa\,\lambda_{\min}(P)}{\|P\|}.$$ (17)

By substituting (17) into (16), one obtains:

$$x^{T}P\varphi(x)\leq\kappa x^{T}Px.$$ (18)

Then, substituting (18) into (15), we have:

$$\frac{d}{dt}x^{T}Px\leq x^{T}\left(\left(G+\kappa I\right)^{T}P+P\left(G+\kappa I\right)\right)x.$$ (19)

Substituting the inequality (19) into Equation (11) leads to:

$$\frac{d}{dt}x^{T}Px\leq -x^{T}Q^{*}x$$
$$\leq -x^{T}\left(Q+K^{T}RK\right)x$$
$$\leq -\left(x^{T}Qx+x^{T}K^{T}RKx\right).$$ (20)
$$\leq -\left(x^{T}Qx+u^{T}Ru\right)$$
$$\leq -L(x,u)$$

From Equation (13) and Equation (20), it is rational and reasonable for us to obtain the equation in Assumption 2.1. This is the end of the proof. The terminal region Ω_{α} is $\left\{\Omega_{\alpha}\mid x^{T}Px\leq\alpha\right\}$.

Definition 1
If there is a system as:

$$x_{k+1}=H(x_{k}).$$ (21)

If there exists a K∞ function $\beta(\cdot)$ making xk subject to $\|xk\|\leq\beta(x0)$, the system will be Input-to-State-Stability (ISS) [35].
Definition 2
If there exists K∞ functions $\chi_{1}(\cdot),\ \chi_{2}(\cdot),\ \chi_{3}(\cdot)$ enabling:

$$\chi_{1}(\|x\|)\leq\Phi(x)\leq\chi_{2}(\|x\|),\Phi(H(x))-\Phi(x)\leq-\chi_{3}(\|x\|).$$ (22)

then the continuous function $\Phi(\cdot)$ is an ISS Lyapunov function of system (21).

Deduction 1
If there exist an ISS Lyapunov function of system (21), the system will be an ISS system.
Theorem 2

If the predictive controllers are calculated in accordance with Equations (5) and (7), the system (2) will be an ISS system.
Proof

The predicted values based on x_{k-1} are denoted as $\tilde{\mathbf{x}}_{k-1+i}$, the corresponding predictive controllers are $\tilde{\mathbf{u}}_{k-1+i}$, and the performance index is J˜; similarly, the predicted values based on x_k are $\hat{\mathbf{x}}_{k+i}$, the predictive controllers are uˆk+i, and the performance index is Jˆ.

Then, (in the following process, Assumption 2.1 is applied):

$$\Delta J = \hat{J} - \tilde{J} = \sum_{i=0}^{N-1}\left(\hat{\mathbf{x}}_{k+i}^T Q\hat{\mathbf{x}}_{k+i} + \hat{\mathbf{u}}_{k+i}^T R\hat{\mathbf{u}}_{k+i}\right) + \hat{\mathbf{x}}_N^T P\hat{\mathbf{x}}_N$$

$$- \sum_{i=0}^{N-1}\left(\tilde{\mathbf{x}}_{k+i-1}^T Q\tilde{\mathbf{x}}_{k+i-1} + \tilde{\mathbf{u}}_{k+i-1}^T R\tilde{\mathbf{u}}_{k+i-1}\right) - \tilde{\mathbf{x}}_{N-1}^T P\tilde{\mathbf{x}}_{N-1}$$

$$= \sum_{i=0}^{N-2}\left(\hat{\mathbf{x}}_{k+i}^T Q\hat{\mathbf{x}}_{k+i} + \hat{\mathbf{u}}_{k+i}^T R\hat{\mathbf{u}}_{k+i} - \tilde{\mathbf{x}}_{k+i}^T Q\tilde{\mathbf{x}}_{k+i} - \tilde{\mathbf{u}}_{k+i}^T R\tilde{\mathbf{u}}_{k+i}\right) \quad (23)$$

$$+ \hat{\mathbf{x}}_{N-1}^T Q\hat{\mathbf{x}}_{N-1} + \hat{\mathbf{u}}_{N-1}^T R\hat{\mathbf{u}}_{N-1} - \tilde{\mathbf{x}}_{k-1}^T Q\tilde{\mathbf{x}}_{k-1} - \tilde{\mathbf{u}}_{k-1}^T R\tilde{\mathbf{u}}_{k-1}$$

$$+ \hat{\mathbf{x}}_N^T P\hat{\mathbf{x}}_N - \tilde{\mathbf{x}}_{N-1}^T P\tilde{\mathbf{x}}_{N-1}$$

$$\leq -\tilde{\mathbf{x}}_{k-1}^T Q\tilde{\mathbf{x}}_{k-1} - \tilde{\mathbf{u}}_{k-1}^T R\tilde{\mathbf{u}}_{k-1}$$

In accordance with Assumption 1.3 and Equation (23), we have:

$$\Delta J \leq -L\left(\mathbf{x}_{k-1},\mathbf{u}_{k-1}\right) \leq -\delta\left\|\left(\mathbf{x}_{k-1},\mathbf{u}_{k-1}\right)\right\|. \quad (24)$$

From Definition 2 and Equation (24), the performance index J is the Lyapunov function of the controlled system (2) based on the control method in Equation (5). Therefore, according to Deduction 1, the system (2) with predictive controllers is an ISS system. This is the end of the proof.

The predictive control procedures are separated into two main parts: the offline part and online part. In the offline part process, the model of predictive control including positive-definite matrixes Q, and R, the feedback controller K, and the positive-definite matrix P serving for solving the terminal region can be calculated. Since the controller uk relates to the latest states xkin each step, we can sensibly obtain the controllers in this part:

$$\mathbf{u}_k = \upsilon(\mathbf{x}_k) \tag{25}$$

In terms of the online part, the states of system (2) will be iterated. Because the predictive controller (25) has been solved offline, we just need to substitute the numeric values xk into Equation (25) to get the values of the controllers in each step. As a result, system (2) can respond rapidly. This is very important for two reasons. On one hand, if the controllers fail to be calculated in a short time, the states of the system will change to xk+τ, where τ is the time for computers to calculate the controllers. The previous controllers (25) cannot enable system (2) to be stable in xk+τwith a large τ. On the other hand, the primary mission of the controller is to maintain a stable system in some special situations. For example, in an electrical system, when the states of the system dramatically change caused by the cut out faulty lines because of the bad weather or the other reasons, the controllers are needed to stabilize the system with the rapid altered states; the nearby connected power plants of the lines should also be controlled in a short time. In these cases, the controllers are supposed to respond as fast as possible to make sure that the system operates stably and safely. Therefore, that is the reason why it is of significance for a system to enjoy rapid responsive controllers.

3. THE MODEL OF A HYDROPOWER PLANT

In this part, the nonlinear model of a hydropower plant is presented. The hydro-turbine, the penstock system, the generator system, and the hydraulic servo system are considered in this model [36]. The relationship between the deviation of the incremental torque and output power can be expressed as:

$$P_m = m_t + \omega , \tag{26}$$

where mt is the deviation of the incremental torque; ω is the variation of the speed of the generator. A typical diagram of the hydro-turbine and penstock system is shown in Figure 1 [37].

We suppose that the cross-sectional area of the penstock is constant. Thus the transfer function of the hydro-turbine and penstock system can be written as:

$$G_t(s) = e_y \frac{1 + eG_h(s)}{1 - e_{qh}G_h(s)} , \tag{27}$$

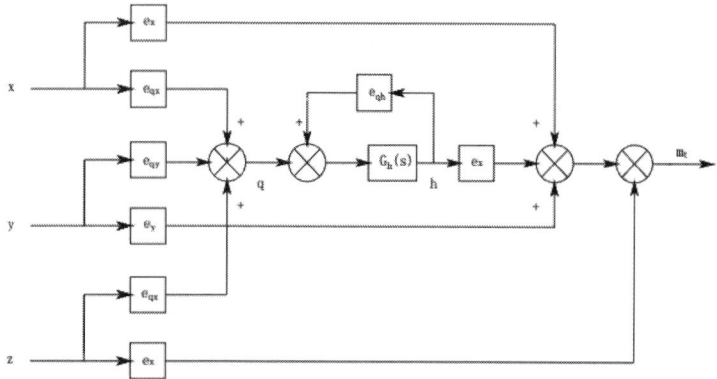

Figure 1. The linear model of the hydro-turbine and penstock system.

where eqh is the first-order partial derivative value of flow rate with respect to water head; e is the intermediate variable; ey is the first-order partial derivative value of torque with respect to wicket gate; Gh(s) is the water hammer transfer function, described by:

$$G_h(s) = \frac{H_A(s)}{Q_A(s)} = -2h_w th(0.5T_r s),$$ (28)

where hw is the characteristic coefficient of the penstock; Tr is the length of the phase of the water hammer wave. From [37], the transfer function of the penstock system can be rewritten as:

$$G_h(s) = -2h_w \frac{\frac{1}{48}T_r^3 s^3 + \frac{1}{2}T_r s}{\frac{1}{8}T_r^3 s^2 + 1}.$$ (29)

Substituting Equation (29) into Equation (27), the transfer function between the incremental deviation of the guide vane opening y and the deviation of the incremental torque mt can be rewritten as:

$$G_t(s) = -\frac{e_y}{e_{qh}} \frac{es^3 - \frac{3}{h_w T_r}s^3 + \frac{24e}{T_r^2}s - \frac{24}{h_w T_r^3}}{s^3 + \frac{3}{e_{qh}h_w T_r}s^3 + \frac{24}{T_r^2}s + \frac{24}{e_{qh}h_w T_r^3}}.$$ (30)

From Equation (30), the state space equations of the hydro-turbine and penstock system can be described as:

$$\begin{cases} \dot{x}_1 = x_2 \\ \dot{x}_2 = x_3 \\ \dot{x}_3 = -a_0 x_1 - a_1 x_2 - a_2 x_3 + y \end{cases}$$

(31)

and:

$$m_t = b_3 y + (b_0 - a_0 b_3)x_1 + (b_1 - a_1 b_3)x_2 + (b_2 - a_2 b_3)x_3,$$

(32)

where x_1, x_2 and x_3 are   state   variables

$$a_0 = \frac{24}{e_{qh} h_w T_r^3}, \quad a_1 = \frac{24}{T_r^2}, \quad a_2 = \frac{3}{e_{qh} h_w T_r}, \quad b_0 = \frac{24 e_y}{e_{qh} h_w T_r^3}, \quad b_1 = -\frac{24 e e_y}{e_{qh} T_r^2}, \quad b_2 = \frac{3 e_y}{e_{qh} h_w T_r} \quad \text{and} \quad b_3 = -\frac{e e_y}{e_{qh}}.$$

A second-order mathematical model of the generator is:

$$\begin{cases} \dot{\delta} = \omega_0 \omega \\ \dot{\omega} = \frac{1}{T_{ab}}[m_t - m_e - D\omega] + u_\omega \end{cases}$$

(33)

where $u\omega$ is predictive controller designed later.

If the influence of the rotor speed on the torque is added to the damping factor, the torque of the electrical load and the terminal active power are equal to each other, *i.e.*,

$$m_e = P_e.$$

(34)

For the generator, the terminal active power can be described as:

$$P_e = \frac{E_q' V_s}{x_{d\Sigma}} \sin\delta + \frac{V_s^2}{2} \frac{x_{d\Sigma}' - x_{q\Sigma}}{x_{d\Sigma}' x_{q\Sigma}} \sin 2\delta$$

(35)

and:

$$\begin{cases} x'_{d\Sigma} = x'_d + x_T + \dfrac{1}{2}x_L \\ \\ x_{q\Sigma} = x_q + x_T + \dfrac{1}{2}x_L \end{cases}, \tag{36}$$

where E'q is the transient internal voltage of the armature; Vs is the bus voltage at infinity; x'd is the direct axis transient reactance; xq is the quartered axis reactance; xT is the short circuit reactance of the transformer; xL is the reactance of the electric transmission line.

The dynamic characteristics of a hydraulic servo system [38] can be obtained as:

$$T_y \frac{dy}{dt} + y = u_y, \tag{37}$$

where y is the incremental deviation of the guide vane opening.

From Equation (26) to Equation (37), combining every part of the governing system into an organic whole as:

$$\begin{cases} \dot{x}_1 = x_2 \\ \dot{x}_2 = x_3 \\ \dot{x}_3 = -a_0 x_1 - a_1 x_2 - a_2 x_3 + y \\ \dot{\delta} = \omega_0 \omega \\ \dot{\omega} = \dfrac{1}{T_{ab}}[m_t - \dfrac{E'_q V_s}{x'_{d\Sigma}}\sin\delta - \dfrac{V_s^2}{2}\dfrac{x'_{d\Sigma} - x_{q\Sigma}}{x'_{d\Sigma}x_{q\Sigma}}\sin 2\delta - D\omega] + u_\omega \\ \dot{y} = \dfrac{1}{T_y}(-y + u_y) \end{cases}, \tag{38}$$

where uy and uω are selected as predictive controllers. The parameters in this paper are w0 = 314, T_{ab} = 8.0, D = 0.5, E'_q = 1.35, $x'_{d\Sigma}$ = 1.15, $x'_{q\Sigma}$ e_y = 1.0, e = 0.7, T_r = 1.0, h_w = 2.0, r = 0, a_0 = 24, a_2 = 3, b_0 = 24, b_1 = 33.6, b_2 = 3, and b3=−1.4, respectively.

4. MAIN RESULTS

The numerical experiments of the nonlinear predictive control of the six-dimensional hydropower plant are presented in this part. In Equation (38), u=(uy,uω)T are predictive controllers designed in accordance with

Equations (5) and (6) later. The following positive-definite matrixes are given as:

$$Q = \frac{1}{2} I_6 = \frac{1}{2} \begin{bmatrix} 1 & & & & & \\ & 1 & & & & \\ & & 1 & & & \\ & & & 1 & & \\ & & & & 1 & \\ & & & & & 1 \end{bmatrix}, \text{ and } R = I_2 = \begin{bmatrix} 1 & \\ & 1 \end{bmatrix}. \tag{39}$$

From what has been mentioned in Equation (8), we have:

$$A = \begin{bmatrix} 0 & 1 & 0 & 0 & 0 & 0 \\ 0 & 0 & 1 & 0 & 0 & 0 \\ -24 & -24 & -3 & 0 & 0 & 1 \\ 0 & 0 & 0 & 0 & 314 & 0 \\ 5.95 & 0 & 0.9 & -0.1228 & -0.0625 & -0.175 \\ 0 & 0 & 0 & 0 & 0 & -10 \end{bmatrix}, \text{ and } B = \begin{bmatrix} 0 & 0 \\ 0 & 0 \\ 0 & 0 \\ 0 & 0 \\ 1 & 0 \\ 0 & 10 \end{bmatrix}. \tag{40}$$

The feedback gain G in Equation (9) is $G = -\frac{1}{2} I_4$, so it is easy to get the feedback controllersK by solving the equation $G=A+BK$, and K is:

$$K = \begin{bmatrix} -5.95 & 0 & -0.9 & 0.1228 & -0.4375 & 0.175 \\ 0 & 0 & 0 & 0 & 0 & 0.95 \end{bmatrix}. \tag{41}$$

By solving the Lyapunov Equation (11) when $\kappa=0.45<0.5=-\lambda_{max}(G)$, the terminal penalty matrix P is:

$$P = \begin{bmatrix} 359 & 0 & 53.55 & -7.309 & 26.03 & -10.41 \\ 0 & 5 & 0 & 0 & 0 & 0 \\ 53.5 & 0 & 13.1 & -1.106 & 3.938 & -1.575 \\ -7.309 & 0 & -1.106 & 5.151 & -0.5375 & 0.215 \\ 26.03 & 0 & 3.938 & -0.5375 & 6.914 & -0.7656 \\ -10.41 & 0 & -1.575 & 0.215 & -0.7656 & 14.33 \end{bmatrix}. \tag{42}$$

Now, after discretizing the system (38), and calculating the predictive controllers based on the above parameters, Equations (5) and (6), the states of the δ and ω of the controlled system, shown in Figure 2, can be obtained by iterating Equation (7).

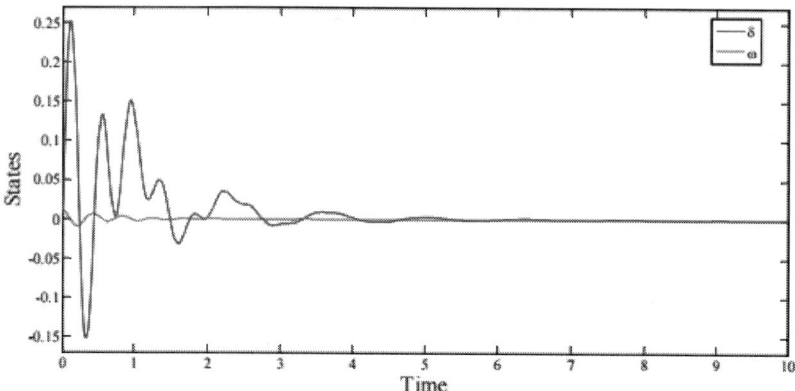

Figure 2. The response of the system (38) with NMPC controllers.

The predictive controllers uy and uω are shown in Figure 3 and Figure 4, respectively.

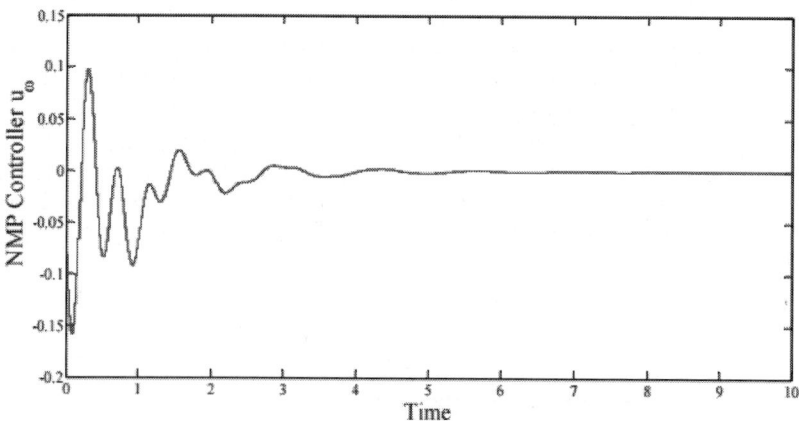

Figure 3. The NMPC controller uω of system (38).

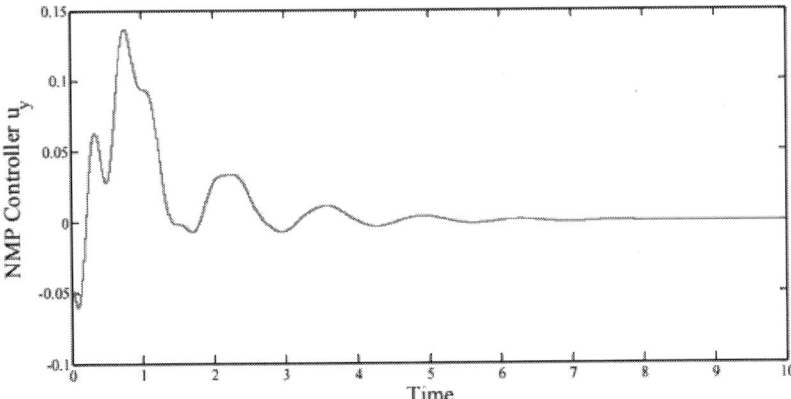

Figure 4. The NMPC controller uy of system (38).

From Figure 2, Figure 3 and Figure 4, it is clear that the states of the hydropower system and the controllers are stable. The computational advantages of the NMPC method used in this paper are significant because the controllers were calculated offline. It should mention that our NMPC algorithm is separated into two parts. Obviously, only the elapsed CPU time of the online part relates to the fast response performance. The computational statistics of the offline part and online part are shown as follows.

Table 1 shows that the elapsed CPU time of the online part involves the entire horizon. We cannot ignore that the CPU used in the experiment is an Intel(R) Core(TM) i5 CPU, M 520 @ 2.40 GHz (4 CPUs), the original mobile version of the Intel(R) Core(TM) i-series. If people use the latest CPU, they will obtain a better performance.

Table 1. The computational statistics of the NMPC method.

Elapsed CPU Time (ms)	
Offline part	5370
Online part	11,569

The values of the performance index are presented as follows. In Table 2, the performance index is presented with gaps of 10 steps. It would be a long sheet if we were to present all the performance indexes here. The performance indexes have little difference when the step is bigger than 191. Thus, we omit the remaining indexes here. From Table 2, we can rationally and reasonably conclude that the performance index tends to zero as the step increases, which means the system is asymptotic stable, corresponding with the theoretic knowledge and the numerical experiment.

Table 2. The performance metrics of the NMPC method.

Number of Step	Performance Index	Number of Step	Performance Index
1	0.117856	101	0.0022
11	0.039339	111	0.012638
21	0.030683	121	0.006103
31	0.050499	131	0.001474
41	0.077429	141	0.000687
51	0.119509	151	0.001066
61	0.016413	161	0.00026
71	0.005165	171	0.000517
81	0.010286	181	0.001085
91	0.002602	191	0.000607

In accordance with the real operational conditions, the issue of the set-point tracking is discussed as follows. The tracking target sets in δ are $R\delta=0.75$, $t\geq6$. The results of the tracking issue are shown in Figure 5. From the figure, we notice that the system has successfully tracked the target and subsequently stabilized the target in a period under the action of the NMPC method.

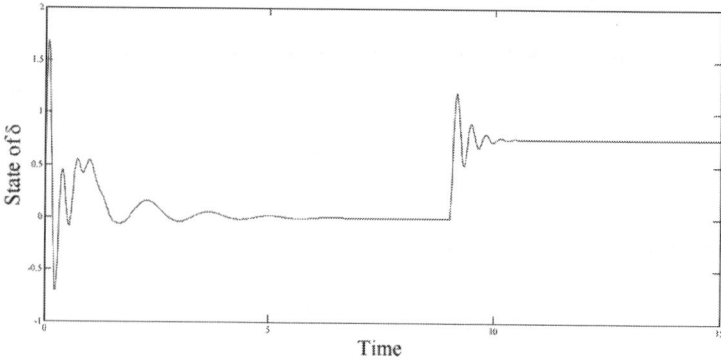

Figure 5. Response of δ of the system controlled by NMPC method to track the set-point.

In several situations, people cannot access the accurate values of the real plant system. In this case, the model-plant mismatch should be considered to verify the performance of the NMPC method in that situation. The results to verify the stability of the proposed method with the mismatches $d\delta=0.05$, $d\omega=0.1$ and $dy=0.02$ are shown in Figure 6.

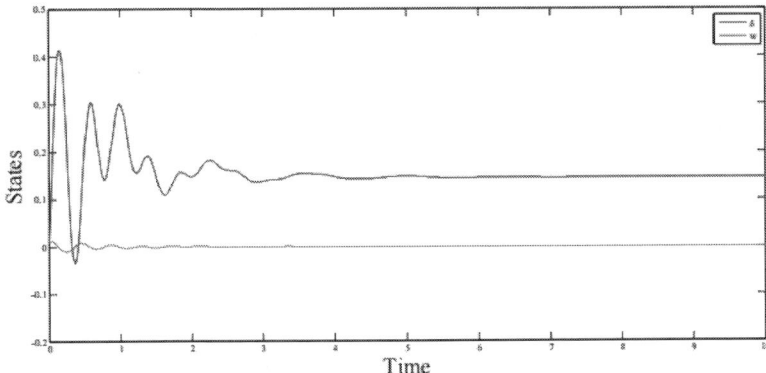

Figure 6. Results of the hydropower station system controlled by NMPC with model-plant mismatch.

Figure 6 shows that the system is still stable with the NMPC controller when there are mismatches existing in the model. The rotor angle δ fails to reach its equilibrium point, whereas it still operates in the stable state. Therefore, although there is a tiny gap between the equilibrium point and the stability state, the system does operate stably. A disturbance is considered here to illustrate the effectiveness of the proposed method when disturbances exist in the system. The disturbance in the numerical experiments exists in δ, which is $\Delta\delta=1$, $10\leq t\leq12.34$. The response of the system in such a disturbance is shown in Figure 7.

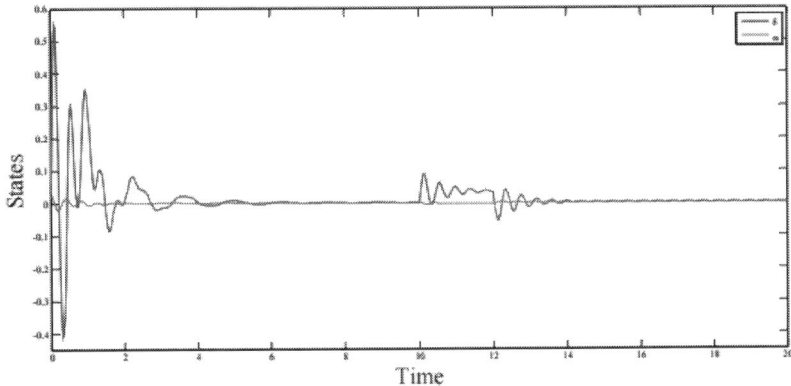

Figure 7. The response of the system with the disturbance Δδcontrolled by NMPC method.

From Figure 7, after a period, the system reaches its steady state, pointing out that the NMPC method enjoys the capacity to make the disturbed system stable. As we know, in a real system, randomness exists in some parts caused by the natural features such as the unpredictable water flow and so on. These unpredictable parts can dramatically damage the system. In this case, the stochastic factors should not be ignored in a control system. The following numerical experiment shows the effectiveness of the NMPC method when there are stochastic factors existing in the system. The results are shown in Figure 8. The stochastic factor is added in the rotor angle δ.

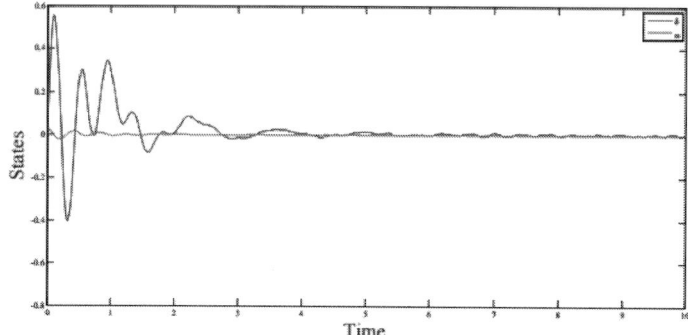

Figure 8. The response of the system with the stochastic factor controlled by the NMPC method.

In Figure 8, the system reaches its steady state in the end. Although, the system spends more time reaching the steady state than the system with the stochastic factor, it is the NMPC method that stabilizes the system to the steady state. That shows the robustness of the NMPC method.

5. DISCUSSION AND CONCLUSIONS

5.1. Discussion

We will further discuss NMPC theory and its applications in other engineering fields in this part. Not only can the NMPC method stabilize the hydropower system introduced in this paper, but it can also be applied in other systems described by state-space equations. More importantly, this control method enjoys the capacity to cooperate with other sophisticated control methods. Those control methods have been used in a variety of fields for many years because of its simple model and the experience. Cooperating with those well-developed methods, the mixed predictive control method is able to take advantages of both sides.

Remark 1
The NMPC controllers enjoy a great advantage in controlling a nonlinear system compared with other traditional control methods such as PID method and so on. The NMPC controllers make it possible for us to control a nonlinear system to the expect states in a short time with less overshoot, which cannot be achieved by PID controllers. In order to illustrate the

effectiveness of the NMPC and PID methods, we introduce the PID controller in the system (38) as:

$$
\begin{cases}
\dot{x}_1 = x_2 \\
\dot{x}_2 = x_3 \\
\dot{x}_3 = -a_0 x_1 - a_1 x_2 - a_2 x_3 + y \\
\dot{\delta} = \omega_0 \omega \\
\dot{\omega} = \dfrac{1}{T_{ab}}[m_i - \dfrac{E_q' V_s}{x_{d\Sigma}}\sin\delta - \dfrac{V_s^2}{2}\dfrac{x_{d\Sigma}' - x_{q\Sigma}}{x_{d\Sigma}' x_{q\Sigma}}\sin 2\delta - D\omega] \\
\dot{y} = \dfrac{1}{T_y}(-k_p(r-\omega) - \dfrac{k_i}{\omega_0}\delta - k_d\dot{\omega} - y)
\end{cases}
\tag{43}
$$

where kp,ki and kd are   PID   parameters,   which are kp=2.0, ki=1.0, and kd=2.0,   respectively [36].   PID   controller   was introduced in state ω.

In order to clearly show the difference between NMPC and PID, we define the index E as:

$$
E = \sqrt{\sum_{i=1}^{n}(y_{0i} - x_i)^2}
\tag{44}
$$

The numerical experiments of NMPC and PID are shown in Figure 9.

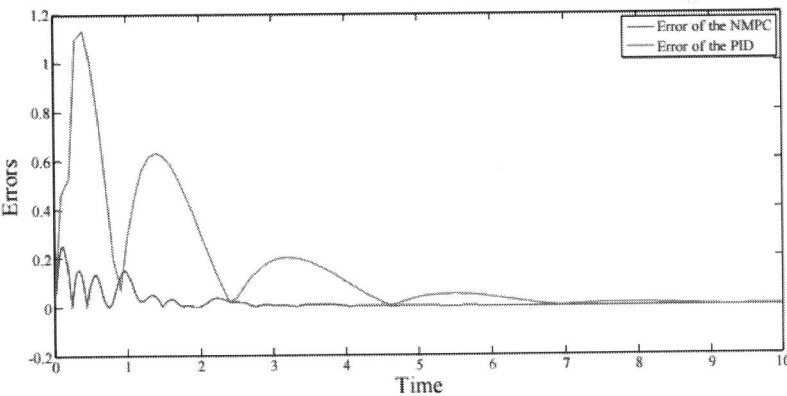

Figure 9. Errors of the NMPC and PID controllers.

From Figure 9, the PID controller stabilizes the nonlinear system (43) slower than NMPC controllers do. Thus, NMPC method has the better capacity to handle a nonlinear system in a short time. It is also shown in Figure 9 that the errors between the expected states and the real states of

PID method are larger than that of NMPC, meaning that NMPC method enjoys an excellent performance in control of the nonlinear hydropower system.

Remark 2

There are several power plants using a simple linear model in control system. The proposed method, of course, is also effective for a linear system. The state-space equations of linear system is:

$$\dot{\mathbf{x}} = A\mathbf{x} + B\mathbf{u}, \tag{45}$$

where u are predictive controllers, A is a n×n matrix and B is n×m matrix.

In order to apply the predictive control method, we need to substitute Equation (37) for the system (1), and discretize the system (45) to iterate. In the process of calculating the terminal penalty matrix P, there is no need to linearize the system through the Jacobian method. One can directly use the matrixes A and B to solve the Equation (11). The linear predictive controller can be obtained from the discretized system (45).

Remark 3

In Section 4, we suppose the system should be controlled at the origin point, but in many other systems, the system is always stable in a given state y0. In this case, the system Equation (1) is:

$$\begin{cases} \dot{\mathbf{x}} = \mathbf{F}(\mathbf{x}, \mathbf{u}) \\ \mathbf{y} = C\mathbf{x} - \mathbf{y}_0 \end{cases}, \tag{46}$$

where y0 ∈ R_l are the expect outputs of the system, y∈Rl are the outputs of the system, and Cis a l×n matrix.

After discretizing the system (46), we should substitute the discretized y for the discretized x in Equation (5), Equation (25) and their sub-equations to control the system to expected states y0. It is to be observed that this is also true for a linear system. Under the circumstance, Equation (46) is:

$$\begin{cases} \dot{\mathbf{x}} = A\mathbf{x} + B\mathbf{u} \\ \mathbf{y} = C\mathbf{x} - \mathbf{y}_0 \end{cases}, \tag{47}$$

and in accordance with Remark 1, system (47) can be controlled to the expected states y0.

Remark 4

For Remark 1, we know that the NMPC method enjoys a great number of advantages compared to the PID method, especially in a nonlinear system. However, these traditional control methods have been used for a long time.

People have built a great body of knowledge about those control methods. For example, PID controllers prevail in almost all the fields, because of their simple structure, stability, convenience and feasibility to set the control parameters. In this case, although NMPC has advantages over the traditional methods, they are not replacing those tradition control methods dramatically. Seldom are the advanced control methods used in practical systems because of the little experience about how to launch such an advanced control method in practice. In addition, the parameters in the advanced control method are more difficult to select than traditional methods. Therefore, we should rationally combine the advantages of the PID with the advantages of other state-of-the-art control methods, for example the NMPC method in this paper. Fortunately, it is possible for us to combine those two control methods.

Take the combination of PID and NMPC methods for example. In order to provide a general method to combine the NMPC method with other control methods, we will generally select PID parameters as the predictive controllers, which means that, in each step, the predictive controller will calculate appropriate PID parameters for the system. We introduce the PID controllers in a nonlinear system (1) as:

$$\begin{cases} \dot{\mathbf{x}} = \mathbf{F}(\mathbf{x}, \mathbf{u}) \\ \mathbf{u} = K_p \dot{\varepsilon} + K_i \varepsilon + K_d \ddot{\varepsilon} \end{cases} \tag{48}$$

where $\varepsilon \in Rl$ and $\varepsilon' = Cx - y0$ are the errors of the system, respectively; $Kp \in Rl, Ki \in Rl$ and $Kd \in Rl$ are PID parameters, respectively. In this case, we have:

$$\begin{cases} \varepsilon = \int_0^\infty (C\mathbf{x} - \mathbf{y}_0) dt \\ \dot{\varepsilon} = C\mathbf{x} - \mathbf{y}_0 \\ \ddot{\varepsilon} = C\dot{\mathbf{x}} \end{cases} \tag{49}$$

Substituting Equation (49) into Equation (48), the PID controlled system is:

$$\begin{cases} \dot{\mathbf{x}} = \mathbf{F}(\mathbf{x}, K_p(C\mathbf{x} - \mathbf{y}_0) + K_i\varepsilon + K_d C\dot{\mathbf{x}}) \\ \dot{\varepsilon} = C\mathbf{x} - \mathbf{y}_0 \end{cases} \tag{50}$$

Then, the PID parameters are selected as NMPC controllers calculated based on the method proposed in this paper. Subsequently, we can iterate the discretized Equation (50) and apply the predictive control method to Equation (50). Undoubtedly, as we mentioned in Remark 2, this strategy can also be applied in a linear system. What we should do is just to introduce a linear model in Equation (48) shown as:

$$\begin{cases} \dot{\mathbf{x}} = A\mathbf{x} + B\mathbf{u} \\ \mathbf{u} = K_p\dot{\boldsymbol{\varepsilon}} + K_i\boldsymbol{\varepsilon} + K_d\ddot{\boldsymbol{\varepsilon}} \end{cases} \tag{51}$$

Substituting Equation (49) into Equation (51), the PID controlled system of a linear system is:

$$\begin{cases} \dot{\mathbf{x}} = A\mathbf{x} + B\left(K_p\left(C\mathbf{x} - \mathbf{y}_0\right) + K_i\boldsymbol{\varepsilon} + K_d C\dot{\mathbf{x}}\right) \\ \dot{\boldsymbol{\varepsilon}} = C\mathbf{x} - \mathbf{y}_0 \end{cases} \tag{52}$$

System (52) can be controlled based on the NMPC method by selecting the PID parameters as the predictive controllers, which means the parameters will be calculated in each step to control the system (45).

Ae numerical experiment demonstrating the proposed combination method is presented here. The system of the numerical experiment is Equation (43). The results are shown in Figure 10. In the numerical experiment, the PID parameters were decided through NMPC method. This has a great advantage. As aforementioned, there are large numbers of PID-based control systems. For improving these systems, people just need to change the method to determine the PID parameters, which is more convenient and economical than replacing the whole system.

From Figure 10, we know that the system is stable, pointing out the efficiency of the method. This is just an example of the combination of PID method and NMPC method or MPC method. Practically, people can combine any given controlled system with the NMPC method or MPC method like this, only if the system can be described as Equation (1) or Equation (45).

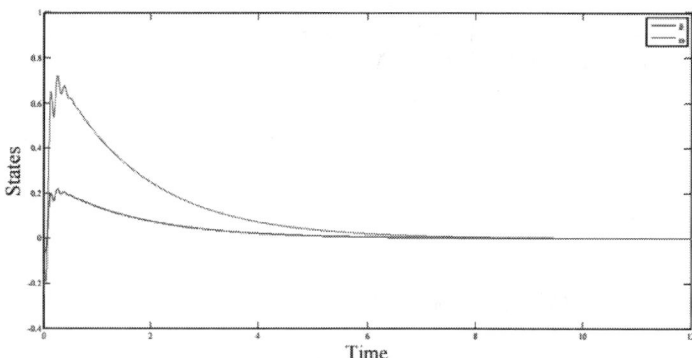

Figure 10. The response of the hydropower station system with PID combined NMPC method.

CONCLUSIONS

In this paper, we propose a simple method to select an appropriate performance index to control a system based on the nonlinear predictive control method. The method is through the use of the Lyapunov equation to select a terminal penalty function, and its stability is proved by the Lyapunov function. Subsequently, a nonlinear hydropower plant system is controlled by the method. The six-dimensional model is complex enough to describe a hydropower system accurately. In the hydropower system, the hydro-turbine system, the penstock system, the generator system, and the hydraulic servo system are considered. Eventually, not only do we compare NMPC method with the traditional PID method, but we also provide strategies to apply the NMPC method to a linear system, and to combine NMPC method with other traditional control methods. It is worth mentioning that the paper not just provides an example of the NMPC method, but offers a general way to use the NMPC method in different situations.

ACKNOWLEDGMENTS

This work was supported by the scientific research foundation of National Science Foundation (51479173, 51279167), Fundamental Research Funds for the Central Universities (201304030577), Scientific research funds of Northwest A&F University (2013BSJJ095), the scientific research foundation on water engineering of Shaanxi Province (2013slkj-12) and the Science Fund for Excellent Young Scholars from Northwest A&F University.

AUTHOR CONTRIBUTIONS

Diyi Chen and Runfan Zhang conceived and designed the experiments; Runfan Zhang performed the experiments; Diyi Chen and Xiaoyi Ma analyzed the data; Runfan Zhang wrote the paper.

REFERENCES

1. Edward, G.R.; Xu, J.C. Mekong hydropower development. *Science* **2011**, *332*, 178–179.
2. Mahmoud, M.; Dutton, K.; Denman, M. Design and simulation of a nonlinear fuzzy controller for a hydropower plant. *Electr. Power Syst. Res.* **2005**, *73*, 87–99.
3. Li, C.L.; Zhou, J.Z.; Ouyang, S.; Ding, X.L.; Chen, L. Improved decomposition-coordination and discrete differential dynamic programming for optimization of large-scale hydropower system. *Energ. Convers. Manag.* **2014**, *84*, 363–373.

4. Martínez-Lucas, G.; Sarasúa, J.I.; Sánchez-Fernández, J.Á.; Wilhelmi, J.R. Power-frequency control of hydropower plants with long penstocks in isolated systems with wind generation.*Renew. Energy* **2015**, *83*, 245–255.

5. Márquez, J.L.; Molina, M.G.; Pacas, J.M. Dynamic modeling, simulation and control design of an advanced micro-hydro power plant for distributed generation applications. *Int. J. Hydrog. Energy* **2010**, *35*, 5772–5777.

6. Mahmoud, M.; Dutton, K.; Denman, M. Dynamical modelling and simulation of a cascaded reserevoirs hydropower plant. *Electr. Power Syst. Res.* **2004**, *70*, 129–139.

7. Zoby, M.R.G.; Yanagihara, J.I. Primary control system and stability analysis of a hydropower plant. In *Power Plants and Power Systems Control 2006*; Westwick, D., Ed.; Elsevier: Kidlington, UK; Burlington, MA, USA, 2007; pp. 165–170.

8. Ren, M.; Wu, D.; Zhang, J.; Jiang, M. Minimum entropy-based cascade control for governing hydroelectric turbines. *Entropy* **2014**, *16*, 3136–3148.

9. Ying, H. Theory and application of a novel fuzzy PID controller using a simplified Takagi-Sugeno rule scheme. *Inf. Sci.* **2000**, *123*, 281–293.

10. Clarke, D.W. Adaptive predictive control. *Annu. Rev. Control* **1996**, *20*, 83–94. Miller, D.E. A new approach to adaptive control: No nonlinearities. *Syst. Control Lett.* **2003**,*49*, 67–79.

11. Mei, R.; Chen, M.; Guo, W.W. Robust adaptive control scheme for optical tracking telescopes with unknown disturbances. *Opt. Int. J. Light Electron Opt.* **2015**, *126*, 1185–1190.

12. Ren, M.; Zhang, J.; Jiang, M.; Tian, Y.; Hou, G. Statistical information based single neuron adaptive control for non-gaussian stochastic systems. *Entropy* **2012**, *14*, 1154–1164.

13. Hu, B.; Michel, A.N. Robustness analysis of digital feedback control systems with time-varying sampling periods. *J. Franklin Inst.* **2000**, *337*, 117–130.

14. Stich, M.; Beta, C. Control of pattern formation by time-delay feedback with global and local contributions. *Physica D* **2010**, *239*, 1681–1691.

15. Xin, B.G.; Wu, Z.H. Projective synchronization of chaotic discrete dynamical systems via linear state error feedback control. *Entropy* **2015**, *17*, 2677–2687.

16. Bidarvatan, M.; Shahbakhti, M.; Jazayeri, S.A.; Koch, C.R. Cycle-to-cycle modeling and sliding mode control of blended-fuel HCCI engine. *Control Eng. Pract.* **2014**, *24*, 79–91.

17. Tian, X.M.; Fei, S.M. Robust control of a class of uncertain fractional-order chaotic systems with input nonlinearity via an adaptive sliding mode technique. *Entropy* **2014**, *16*, 729–746.

18. Yashar, T.; Wang, J.D. Chaos control and synchronization of a hyperchaotic Zhou system by integral sliding mode control. *Entropy* **2014**, *16*, 6539–6552.

19. Chen, D.Y.; Zhao, W.L.; Ma, X.Y.; Zhang, R.F. No-chattering sliding mode control chaos in Hindmarsh—Rose neurons with uncertain parameters. *Comput. Math. Appl.* **2011**, *61*, 3161–3171.

20. Long, L.J.; Zhao, J. Adaptive fuzzy tracking control of switched uncertain nonlinear systems with unstable. *Fuzzy Sets Syst.* **2015**, *273*, 49–67.

21. Huang, H.; Chen, L.; Hu, E. A neural network-based multi-zone modelling approach for predictive control system design in commercial buildings. *Energy Build.* **2015**, *97*, 86–97.

22. Chen, H.Y.; Liu, M.Q.; Zhang, S.L. Robust H_∞ finite-time control for discrete Markovian jump systems with disturbances of probabilistic distributions. *Entropy* **2015**, *17*, 346–367.

23. Grüne, L.; Pannek, J. *Nonlinear Model Predictive Control*; Springer: London, UK, 2011.

24. Chen, H.; Allgower, F. A quasi-infnite horizon nonlinear model predictive control scheme with guaranteed stability. *Automatica* **1998**, *34*, 1205–1217.

25. Chen, C.C.; Shaw, L. On receding horizon feedback control. *Automatica* **1982**, *18*, 349–352.

26. Mayne, D.Q. Model predictive control: Recent developments and future promise. *Automatica* **2014**, *50*, 2967–2986.

27. Xu, J.; Huang, X.L.; Mu, X.M.; Wang, S.L. Model predictive control based on adaptive hinging hyperplanes mode. *J. Process Control* **2012**, *22*, 1821–1831.

28. Martí, R.; Lucia, S.; Sarabia, D.; Paulen, R.; Engell, S.; de Prada, C. Improving scenario decomposition algorithms for robust nonlinear model predictive control. *Comput. Chem. Eng.* **2015**, *79*, 30–45.

29. Zeng, X.H.; Yang, N.N.; Wang, J.N.; Song, D.F.; Zhang, N.; Shang, M.L.; Liu, J.X. Predictive-model-based dynamic coordination control strategy for power-split hybrid electric bus. *Mech. Syst. Signal Process.* **2015**, *60*, 785–798.

30. Li, S.Y.; Zhang, Y.; Zhu, Q.M. Nash-optimization enhanced distributed model predictive control applied to the Shell benchmark problem. *Inf. Sci.* **2005**, *170*, 329–349.

31. Yang, J.; Li, X.; Mou, H.G.; Jian, L. Predictive control of solid oxide fuel cell based on an improved Takagi-Sugeno fuzzy model. *J. Power Sources* **2009**, *93*, 699–705.

32. Roubos, J.A.; Mollov, S.; Babuska, R.; Verbruggen, H.B. Fuzzy model-based predictive control using Takagi-Sugeno models. *Int. J. Approx. Reason.* **1999**, *22*, 3–30.

33. Sarimveis, H.; Bafas, G. Fuzzy model predictive control of non-linear processes using genetic algorithms. *Fuzzy Sets Syst.* **2003**, *139*, 59–80.

34. Jiang, Z.P.; Wang, Y. Input-to-state stability for discrete time nonlinear systems. *Automatica* **2001**, *37*, 875–867.

35. Xu, B.B.; Chen, D.Y.; Zhang, H.; Wang, F.F. Modeling and stability analysis of a fractional-order Francis hydro-turbine governing system. *Chaos Soliton. Fract.* **2015**, *75*, 50–61.

36. Shen, Z.Y. *Hydraulic Turbine Reglation*, 3rd ed.; China Water Press: Beijing, China, 1998. (In Chinese)

37. Chen, D.Y.; Ding, C.; Do, Y.H.; Ma, X.Y.; Zhao, H.; Wang, Y.C. Nonlinear dynamic analysis for a Francis hydro-turbine governing system and its control. *J. Franklin Inst.* **2014**, *351*, 4596–4618.

Index